the
vanishing
ice

the
vanishing
ice

Diaries of a Scottish snow hunter

Iain Cameron

Vertebrate Publishing, Sheffield
www.v-publishing.co.uk

the
vanishing
ice

Iain Cameron

First published in 2021 by Vertebrate Publishing.
Hardback edition reprinted in 2021. Paperback edition first published in 2021.

 Vertebrate Publishing
Omega Court, 352 Cemetery Road, Sheffield S11 8FT, United Kingdom.
www.v-publishing.co.uk

Front cover: The snow cathedral of Ben Nevis, September 2016. © Murdo MacLeod.
Author photo: The author partially hidden beneath a snowdrift on Ben Nevis in July 2016. © Alistair Todd.
Other photography by Iain Cameron unless otherwise credited.

This book is a work of non-fiction. The author has stated to the publishers that, except in such minor respects not affecting the substantial accuracy of the work, the contents of the book are true.

A CIP catalogue record for this book is available from the British Library.

ISBN: 978-1-83981-108-1 (Hardback)
ISBN: 978-1-83981-087-9 (Paperback)
ISBN: 978-1-83981-088-6 (Ebook)
ISBN: 978-1-83981-089-3 (Audiobook)

Cover design by Jane Beagley, Vertebrate Publishing.
Production by Cameron Bonser, Vertebrate Publishing.
www.v-publishing.co.uk

Vertebrate Publishing is committed to printing on paper from sustainable sources.

Printed and bound in the UK by TJ Books Limited, Padstow, Cornwall.

To Adam.
Though no longer with us,
his presence is enduring and ubiquitous.

Contents

prologue

In the belly of the beast

I emerged at last, breathless, on to the plateau. The ominous-looking wintry squall that had been chasing me up the hill arrived from the west just as I completed the last few heavy steps of a relentless climb from the col, some 1,200 feet below. As I stood upright once more and filled my lungs with the rarefied Lochaber air, the first outlying snow-flakes of the squall danced down harmlessly enough. The initial flurry tends to do that: a few slivers corkscrew down from the sky and land softly on your sleeve or face, just on the leading edge of the gale that rides invariably behind them. But this is, literally, the calm before the storm. I've been in this situation more times than I can remember, and I know what's coming next.

Overnight snow already lay thick on the plateau, and the forecast promised more for the rest of the day and the next. Taking a minute to orient my internal compass and restore my breathing to something resembling normality, I reached into the side of my rucksack for one of the bottles of flavoured water I carried, only to find that there was ice floating in it. As I tipped my head back to pour the freezing mixture into my mouth, the first wave of the storm hit me – hard.

* * *

The day had started three hours earlier in the deserted glen far below. A man I trusted – a long-time snow observer – had told me that there were two old patches left over from the previous winter in one of the high corries on this hill.[1] This seemed unlikely to me, as the autumn weather had been mild – warm, even. I needed to be sure, though. I *had* to count every last relic of snow from the preceding winter before they were buried for another nine months or so. Missing even one would have felt like failure. But there was a problem: I had never been to this corrie before. It lurked on the edge of my knowledge. An apparently tricky place to reach even in summer, it was going to be doubly so in winter conditions. But, since no one else had volunteered to venture into this neglected corner of the Highlands, I decided there was no alternative but to travel there myself.

The climb from the car park had started memorably, but not for the reasons one might hope. On the upward pull that eventually reaches on to the col, about twenty minutes after starting, I chanced across a dying red deer. He was a handsome old stag and lay about fifteen feet from the stalkers' track. His front-right leg looked badly broken and, clearly, he could not move. His big brown eyes stared at me as I neared him, and he opened his mouth, attempting a roar to scare me off but managing only a hollow, thin rasp. It was a pitiful, tragic sight to see, and I could offer him no solace. I moved off, regretfully, hoping his death would come quickly and that this sad, chance encounter wouldn't amount to a macabre omen for my trip.

About an hour and a half later, I arrived at the foot of the col, where the first impediment presented itself. The overnight snow had been heavier than forecast and my path upwards on to the plateau was unfathomable beneath the fresh falls. All around me the landscape was winter, not autumn. The late-year, lifeless vegetation poked through the snow in an apparent attempt at defiance of its surroundings. Thick and ominous cloud swirled around the higher reaches of the hills to the west, whose summits were coated white.

---

1   A corrie is a glacial basin in the side of a hill, the Scottish Gaelic equivalent of an Alpine 'cirque'.

In the absence of any discernible path, I could do nothing else but one thing: climb. So, picking out the least hazardous and most obvious-looking route, up I went. It was steep. *Really* steep. I gained height quickly as a result, though, and luckily the path became clearer. The snow that had fallen filled the hollows where thousands of boots had trodden, and it snaked up the escarpment at a more forgiving angle – although to call it a path would have been generous. It was no better than an indistinct goat track, and goats would have thought twice about ascending it in these conditions. It became steeper the higher I got, and as I paused half-way up for breath, I turned around to see whence I'd come. The hills that encircled me were splattered with thick snow, and any heather that didn't lie under the white blanket was a dull, lifeless mahogany brown. A dark line of approaching weather, the squall of snow, raced towards me. I reckoned I had fifteen minutes to reach the top before it caught me up and deposited its considerable payload.

But moving upwards in these conditions was not an easy endeavour. The rutted path soon vanished yet again, leaving me to judge the steep slope and where best to place my hands and feet. I cursed myself for not coming a few days previously when the weather had been better.

I made slow progress, but progress nonetheless. At length I managed to negotiate a large bluff and grasped a solid-looking boulder for purchase to heave myself up.[2] To be safe I tested its grip on the ground by means of a stout pull, which was just as well, for it rolled from its berth almost without resistance and disappeared down the hill to the bottom, some 1,000 feet below. My heart beat loudly as I watched it tumble, and the clatter of it smashing off a granite outcrop raced back up the hill and into my ears. In less than ten seconds, the stone had reached the point where I had stood thirty minutes earlier. I cursed myself once again.

But then, I was up. The last big step negotiated, I arrived on the plateau. I was still at least an hour from where the two old lumps of snow from the previous winter were purportedly located. Finding them would not

---

2   A bluff is a steep headland.

be easy. Would they even be visible by now? If they *had* survived, would I know? The new snow might have buried them.

* * *

As I emptied that drink of icy, flavoured water into my mouth, I felt a push on my back as though I'd been hit by a car, and it propelled me forward on to my knees. I laughed a hollow chuckle to myself to try and brush it off, but this was serious. I needed to give the weather the respect it deserved. There didn't seem any point in pretending anything else. To accompany this seemingly tenfold increase in windspeed, the flakes of snow started to fall – if fall is the right word – horizontally. I took a bearing and made for the drop-in point to the corrie on the other side of the hill where the old snow would, hopefully, be lying. But before a minute had passed, I could see less than thirty yards. Another minute later it was half that. 'It'll pass. It'll pass,' I said to myself over and over as I walked across the plateau. By this point the ground and sky had fused together into a maelstrom of choking white. I checked my bearing again and trusted in it. I had to. This weather was as bad as any I'd witnessed on the hills of Scotland.

The walk that day across the plateau, which reaches almost 4,000 feet, can best be described as a penance. The buffeting by the gale was intense, easily above storm force. It once threatened to send me airborne in its ferocity. The snow, too, could scarcely have been thicker. With visibility this bad, it would be all too easy to walk off the edge of a cliff, so I stopped every few minutes to check my location. So far, so good.

But, despite the grim conditions, I made good progress across the plateau, and quicker than I anticipated, doubtless due to the wind paddling me on. At length I saw the rim of the other side of the summit ridge and, to my relief, it now looked as though the falling snow was moderating. I took the map out once more and checked my position. By good fortune, perhaps, I had arrived just near the drop-in point of the corrie. The sky now turned from white back to grey, and the wind eased to a playful breeze as quickly as it had started.

Gazing into the abyss below was like staring into a murky cauldron that contained otherworldly ingredients. Great swirls of snow and cloud spun and danced before me. The tail end of the gale whipped eastward, ready to inflict a sting on the next range of hills in its path. For me, at least for now, the ordeal was over. I sat on the snowy grass and put crampons on my boots for grip and waited a minute to see if the snow would lessen further, which – mercifully – it did. I decided, at last, to drop in.

Very deliberately, I eased down the steep grassy slope which itself wore a substantial covering of snow. The wind that had flattened me on the summit had all but disappeared in the shelter of the corrie. I relaxed my body. The absence of noise lent this place an altogether more serene atmosphere than just twenty yards away on the summit ridge. I breathed easier, too. The snow still fell, though, even if only in light, large flakes. It came up to my ankles and I was glad of the crampons on my feet and the ice axe in my hand, though there wasn't enough snow to anchor on if I should fall, so the axe locked into my hand as a mental crutch and very little else.

In the high places of Lochaber, visibility can be so poor, even during the middle of the day, that it is a struggle to see more than twenty yards ahead if the weather is appalling. The cloud and snow can be so thick that you must trust to navigation and instinct. It's an unsettling experience. Half of your mind – the subconscious – is telling you all the while to get out of this place; the other half – the pragmatic – wills you on to complete your task. I had come too close to the prize to let my subconscious win out.

But then, without warning, the snow started to fall thickly again, and I could hear the wind howl off the top of the cliffs above. Time was now against me. I needed to find last winter's snow if it were there and get out of the corrie before the new snow barred my exit. I had good coordinates, but in this decaying place a compass offered little in the way of help. I instead used instinct and judgment as I traversed along the foot of the cliffs that, on my right-hand side, reached up 300 feet into the vortex of wind and snow. On my left there was nothing except cloud and a steep slope that did not break for some 1,000 feet. Slipping here could not be countenanced. 'Steady, Iain,' I repeated to myself. For ten minutes I tottered across broken rock and drifting snow, hoping that at any second I'd stumble across the icy relics from a year ago. But nothing. 'They've probably gone,' I said to myself, just on the cusp of giving up. Mentally I was almost spent. 'Another few steps. Another thirty seconds.'

Just then, however, the ground changed ahead. 'What's that?' I said out loud, as though someone were with me. I took a few more steps forward. I strained my eyes and wiped my glasses. In front of me, not more now than ten yards distant, there they lay. Two big slabs of dirty white snow sat at the foot of the cliff, almost invisible in the camouflage of their surroundings. In form they resembled large white coffins which, considering the environment I was amid, seemed grimly appropriate. 'YES!' I shouted, as though I'd discovered some long-lost, priceless treasure in one of the deep places of the world. I was ecstatic that they endured and that I had found them. They were larger than I'd expected, too, and as I crouched beside them, I knocked on their surface: hard as the rock that they sat on. Months and months of compression, melting and refreezing had cast them like concrete. Their tops were like dragon skin, scaly and scalloped – and just as tough. There was no doubt that, at about fifteen feet long each, they'd survive now. The snow that had already fallen half buried them, and the heavier snow that was due in the next day would cover them up for the season. This would be the last time I, or anyone else, would see them for many months.

But how had it come to this? The snow that lingered on this hill never used to melt. Or, at least, it hardly ever did. Travellers in previous centuries even commented on being able to see it from the valley below all-year round. 'The only snow visible from a British train station in every month of the year,' one excited journal read in the 1930s. No longer could the same thing be said. The snow was a shadow of its former self, reduced to two small lumps of ignominious irrelevance, hidden from, and of no interest to, all but the most dedicated enthusiast. This was the first year in five that any snow had endured the whole year in the corrie. Could it be a sign of things to come? Had climate change stretched its long fingers into even the smallest nooks of the highest Scottish hills?

I had no time to ponder this any further. The snow was deep, and I had to get out of this cauldron of ice and rotting rock. Fresh boulders from rockfalls lay all around, serving as a reminder that this landscape is constantly evolving and not a place to linger.

I retraced my steps back up the steep slope and on to the lip of the plateau. As I came over the edge, the blast of wind that hit my chest forced me to sink down momentarily and grab the vegetation. The snow stung my face so hard it felt like I was being sandblasted. At that point it was obvious that going back down the way I had come would be tantamount to a death wish, so I abandoned the idea. Though the car lay that way I could not risk the descent back down to the col in such atrocious conditions. Poor decision-making on the hill can cost lives and I had no wish to call mountain rescue, nor to become another statistic. An easier route off the north side of the hill would necessitate a taxi journey back to the car, but it would be money well spent.

I descended slowly, and with every downward step I took I cared less about the weather that chased me off the hill. I was beyond pleased I had come to this place today because, had I not, the two patches would have gone unrecorded. When researchers look back in a hundred years' time, they will maybe remark that in 2007 one enthusiast noted that two small patches of snow had survived the year on this hill. This may just be a footnote in a paper they write, and they will not know the human story that lies behind the record. This is of no moment to me. What matters is the snow that was recorded, not the means.

one

The spark

I got up first, as usual. It was early, about 6.30 a.m. The sun streamed in through a gap in the curtains of our bedroom and, for the first time in what seemed like over a week, I hadn't woken up to rain bouncing off the window. Letting my six-year-old brother sleep on, I went upstairs to the living room to put on the television. Prior to pressing the heavy analogue button, which rocked the TV back on its stand, I opened the curtains to let the light of the day in. There were no clouds in the azure blue sky that morning and on the hills, spring's first flush of growth had turned the grass a verdant green. The warmth of the sun could be felt through the single-paned window. Even then, just nine years old on that early May day in 1983, I was in awe of the views our new house had. It sat right at the top of a housing estate in Port Glasgow, literally the highest street in town, 500 feet above the Firth of Clyde. We'd moved there just a few months back, when my dad started a new job locally. The vistas from the front of the house were panoramic and stretched over the estuary from the Erskine Bridge in the east to Dunoon to the west. The high hills on the opposite side of the Firth looked impossibly far away and seemingly unclimbable. Normally they poked into the clouds and were invisible from the house. Some of the hills were rolling and green, but others were grey and craggy. How could anyone, even a child, not be drawn in by the allure of these new visions?

One hill, many miles distant, towered higher than the others. It rose up from behind two smaller hills and the summit cone sat atop two large shoulders which jutted out horizontally in pleasing symmetry. On that beautiful May morning, sitting on this distant peak's south-facing slope, gleaming prominently in the early sunshine, was a large patch of what seemed to be brilliant-white snow. I couldn't understand how it had endured. I stood for a moment and looked at it. 'It can't be,' I said to myself, doing a double take. No snow had fallen at my house for many weeks. Surely it had all melted? The last remnant I'd seen that year had been a long, streaky drift against a lichen-covered drystone wall, just a hundred yards or so behind our garden. But that was ages ago. Yet here on this muscular, high hill there remained an apparent relic of winter just past, clinging on despite the warm weather. Why, when it had melted away everywhere else, did it persist here? Though old enough to understand that there must be a reason for it, I had not yet enough wit to work out why.

For reasons that I cannot adequately convey, my curiosity had been piqued to the extent that I wanted – needed – to identify which hill's chest this white medallion hung on. To do that I was going to have to conduct my own detective work, as I knew nobody else in my family would be able to help me. They weren't much interested in the outdoors, so any attempt to enlist them as helpers would have been a futile exercise. That said, I remembered that somewhere in the house we had a tatty old AA road atlas that my parents kept, despite not having a car. I unearthed it from the ragged old ottoman that the phone sat on, only to find the atlas in equally poor repair. There was now no front or rear cover, and the staples were fighting a hopeless, rearguard action to hold the whole thing together. It was in miserably large scale, and covered in coffee stains to boot, but I figured it might contain some clues. Even then I was fascinated by maps and geography so, standing with the atlas open at the window, I oriented it in the direction I faced. Glancing between the open atlas and the view, as though trying to crack a secret code, I finally surmised that the hill where the snow reposed could *only* be Ben Lomond. According to the atlas it rose to an impressive 3,194 feet. I knew that people in

Scotland called hills over 3,000 feet 'Munros'. I'd no idea why this should be the case, but I was pretty sure that I was gazing upon Scotland's most southerly Munro.

I do not recall that morning whether I mentioned this revelation to my parents. I suspect not. It was pointless asking them in any case why snow would still be there on Ben Lomond. They'd barely set foot on a hill, never mind studied the vicissitudes of mountain weather. But, that morning, a light bulb flicked on in my young mind. A spark. As for why the snow hung on, that was something I needed to figure out for myself.

For the next few days the pattern remained the same: up at first light, run upstairs and peer through the window to see if the snow was still there, go to school, then look again in the evening before I went to bed. It had turned into a strange fascination. But, at length the weather turned again and the hills were obscured by cloud and rain, so for three or four days I saw nothing. 'Surely it's melted,' I said to myself. Then, one day after returning from school, the cloud blew off Ben Lomond and its cohorts. The snow, alas, had gone. At least, I couldn't see it. I was disappointed by its demise. Sad, even. I knew well enough that it eventually *had* to vanish, but this didn't lessen the feeling of disappointment. It felt like saying goodbye to a relative you'd probably not see again for ages. I was determined that, one day, I would get to the bottom of the mystery and visit this snowy place.

Because of my parents' lack of interest in anything to do with hills, or the general outdoors, and because independent motorised transport was beyond our financial means, my interest in patches of snow remained confined for the next eight years to what I could see through the living room window and from the Ordnance Survey trig point about half a mile from our house (the views from there were even more wideranging and took in hills that were over forty miles distant). The option of what I could glean from very primitive technology presented itself, too. Chief amongst these was page 421 of Ceefax. Every morning in winter I would load up this highly unwieldy and slow proto-internet. Page 421 showed all five of the Scottish ski resorts at that time, and how many of the lifts were running. A teenage me wanted to know how much snow lay on the hills, even though I had never skied in my life, nor climbed a snowy hill. It was completely illogical. I remember at the age of thirteen being castigated by my parents for dialling a premium-rate phone number that gave a real-time overview of which ski lifts were spinning at Scotland's resorts. I think, actually, they were quite relieved when they found out what I had used the phone for, and that their eldest son wasn't calling an altogether less wholesome premium-rate service.

At secondary school I recall disagreeing with my geography teacher on the subject of snow. I didn't rate him, mostly because he used to enjoy demeaning his pupils by showing them how poor their knowledge of his precious subject was. It wasn't just that, though. He seemed disinterested in our learning, often teaching in what I thought was an overly perfunctory manner. I wouldn't have it, and he disliked me from day one. The snow disagreement centred around a road. He said to the class one day that the A939 Cockbridge to Tomintoul had been closed in the past by snow in every calendar month of the year. I knew this to be nonsense (it had never closed in July or August) and told him as much, though obviously not in such blunt terms. It was a nail in my educational coffin. Looking back on episodes like this, I wish I'd learnt to keep my mouth shut.

The school that I attended, St Stephen's High, could often be found propping up every other school in the annual exam results league tables for Strathclyde Regional Council, as it was back then. Many people speak of their time at secondary school as the happiest of their lives, but for me they were the worst and I despised it, as I did every subject bar English and geography. As a result of this disinterest and apathy, I decided to leave school at sixteen with only a handful of mediocre qualifications, much to my parents' chagrin. I had no clear idea of what I wanted to do next, or how to do it; I was just thankful to be out of that awful place. However, by an unusual turn of good fortune, I managed to secure an apprenticeship to be an electrician with a local, family-run company. For the next four years, this provided me with welcome cash, and with the discipline, learning and practical skills that I cherish to this day. In contrast to the gloom of my secondary education, I look upon these four years fondly as some of the best I've ever had. They were the making of me. (Though my siblings and I were brought up with an abundance of love, discipline and a good sense of right and wrong in the home, an apprenticeship among older, skilled men instilled in me a sense of diligence and hard work that can hardly be overstated. The pride with which these ordinary men worked still resonates with me thirty years after I was first exposed to it.)

As soon as I was able to, I learnt to drive and got a car. I wanted to explore my country, something that I had never been able to do independently. I wanted to climb hills and to go to see some snow in summer. However, having lived a relatively sheltered life thus far, I was much too nervous to do anything ambitious. My first walks were done on easy hills with obvious paths and tracks. I recall one such trip, with my brother. Aged seventeen and fourteen, we climbed up Ben More near Crianlarich in early June. Mum and Dad, perhaps blissfully unaware of the potential risks of letting two children loose on the hills unsupervised, or perhaps just trusting in our ability, let us get on with it. Being Britain's sixteenth-highest hill, Ben More felt like a big adventure – a huge and hulking mass of steep grass and rock. Just below the summit, we came across fourteen ptarmigan, which were the first I'd ever seen. I was transfixed and followed them as they deftly scuttled across the rocks, always just too far away for a decent camera shot on the Olympus Trip I had bought from a charity shop. At the summit itself, patches of snow on Ben More's neighbour Stob Binnein were visible. What a treat. A whole world of new possibilities and excitement. I remember it as yesterday.

Around this time, interest in the outdoors was growing, though it was very much a minority pastime. TV programmes appeared which hitherto would have struggled to get a commission. This was to a significant extent the result of a hugely popular TV show called *Weir's Way*, presented by the irrepressible Tom Weir. A Scottish staple during the 1970s and 1980s, *Weir's Way* showcased the bobble-hatted Tom visiting parts of the country that were not well known to the average viewer. Weir remains a legendary figure to the outdoors community in Scotland; a pioneer of opening up the outdoors to a wider audience via his written and TV output.

Of the TV shows that appeared in the early 1990s, the most popular was Muriel Gray's *The Munro Show*. This programme had a big effect on me and many more folk my age. It was a window to view parts of Scotland that I scarcely knew existed, save in the maps that I pored over as a teenager. Gray had an infectious enthusiasm too, which came across well on screen. The format of the show was original and highly engaging. It started with each walk being introduced by the extremely eccentric Gaelic poet Sorley MacLean, sitting on an armchair. Next to him was a blackboard with the hill's name written on it. MacLean then translated the often unpronounceable and impenetrable name, sometimes imparting an anecdote that he'd either heard or experienced himself. Watching as a young man I thought him a bit odd but, viewing the programme back on YouTube many years later, I completely underestimated just how wonderfully over-the-top he was.

Once the meaning and pronunciation had been delivered by MacLean, the rest of the show contained splendid scenic shots of Gray wandering up whichever hill she and her team had decided to climb.

One episode of *The Munro Show* stood out. In it Gray walked up Braeriach on a beautiful late summer's day, to judge by the height of the sun and the colour of the vegetation. This hill is the second-highest in the Cairngorms and third-highest in the UK. As she walked across the vast, Arctic-like plateau, the camera cut to her standing on a precarious rocky spur, with what looked like drops on all sides of hundreds of feet. It then panned downwards into the black abyss, stopping when it framed four large patches of snow.

'In this corrie,' Gray continued, 'is Scotland's most permanent snow-field. In fact, it's only melted twice this century, in 1933 and 1959.'

Upon excitedly hearing this gem of information, I sat bolt upright in my seat. Several things leapt into my mind. Firstly, how did she know they were the most permanent snowfields in Scotland? Secondly, who on earth was monitoring long-lasting patches of snow on Scotland's hills? Finally, why did patches of snow persist all year? This information was a revelation to me. Needless to say, I recorded all the episodes of this show on VHS and would come back time and time again to the Braeriach episode just to see those massive banks of snow in the shaded corrie below. At that time, I thought them impossible to reach, but I longed to see them nonetheless.

Inspired by the revelatory information that people must be monitoring long-lasting snow in Scotland, I vowed to find out more. This was a much taller order in the early 1990s than nowadays, with no internet to draw on. Finding out such things required visiting libraries and bookshops. Just finding the latter in the area where I grew up was hopeless because establishments that sold books were often confined to the larger John Menzies or WHSmith newsagents. Apart from the odd Ordnance Survey map or glossy 'Scotland is Beautiful'-type tartan shortbread books, there was nothing remotely like the sort of tome I yearned for.

The first few sorties to the local public libraries yielded little either, with librarians knowing virtually nothing about the outdoors, never mind this esoteric subject. But, because diligence is the mother of good fortune, one day in a bookshop in Glasgow I came across a publication that gave me my eureka moment.

The outdoor section of Waterstones bookstore in Glasgow wasn't particularly well-frequented at that time. I recall perusing it at leisure, with no one else in close proximity. Because it contained the largest section of its type in Glasgow, I figured it would be the best option for locating any literature that might conceivably cover this odd interest. Sure enough, it was, and I stood rooted to the spot as I read a particular chapter of a book I found. Many people will have experienced a similar sensation when they read something that electrifies them or informs them of something that makes them think, 'Yes! Yes, that's it!' It is a rare occurrence but, because of that, one is inclined to remember it for the rest of one's life.

Martin Moran's *Scotland's Winter Mountains* was a revelation to me.[1] Intended as a guide for mountaineers, its pages were packed full of technical information and instruction on winter climbing. Most of the content didn't interest me. At that point I had little or no desire to indulge in rock

---

1   Moran, Martin (1988), *Scotland's Winter Mountains*, Pynes Hill, Exeter: David & Charles Ltd.

or ice climbing, but what grabbed my attention was a brief section that specifically dealt with snow. The chapter in question discussed the last ice age, many thousands of years ago, and the more recent so-called 'Little Ice Age' of the period 1600–1850 (or thereabouts). On one page it posed the hypothetical question, 'Are today's snow patches dying relics of a bygone era or the harbingers of new glaciers?' This was what I'd been searching for! This question accompanied a full-page table that had some golden nuggets of information, such as a list of hills and long-lying snow patches, the elevation they sat at, their aspect and some commentary on them. These observations on snow patches were wonderful. 'Never known to wholly disappear.' 'Not closely observed but occasional survival has been noted.' It was fascinating. Names that nowadays I am familiar with to the point of saturation, I read for the first time – Ciste Mhearad, Aonach Mòr, Ben Wyvis and many more. Proof in print, at last, that these patches were being monitored. There was also a photograph of the same snow patches that Muriel Gray showed us on the Braeriach episode of *The Munro Show*, with massive drifts visible in a picture from October 1984. I simply had to have this book and would have bought it on the spot, but for one small problem: its price tag.

Even these days, shelling out £25 on a book would make many people think twice. Spending that amount in 1990, when my weekly take-home pay from my apprenticeship was about £50, meant plenty of patience and discipline. So, for the next five weeks I carefully saved £5 each week, setting it aside and putting it behind a poster of Madonna that hung by Blu Tack on my bedroom wall. It meant I couldn't buy any of the nice clothes or fancy trainers that I had developed a penchant for when I started earning my own money. This small sacrifice for the pleasure of what was sure to be a fascinating chronicle seemed worth it. Indeed, it became one of my most prized possessions.

Going back to Waterstones to buy the book was exciting. I even eschewed a gathering of the other apprentice electricians from my college who were going into Greenock to play pool and then on to the amusement arcade. I didn't dare tell them the reason for doing this. To do so would have meant inviting some unwanted adolescent abuse.

The woman who served me that day smiled as I handed over the five £5 notes. I put the book in a shiny Waterstones carrier bag and ran back to the station to catch the train home. I seldom put it down for weeks thereafter.

After finishing my electrical apprenticeship in October 1993, I took some time out and worked, briefly, as a volunteer for the National Trust for Scotland at Glen Coe. In between (unpaid) dogsbody duties such as red deer counting and digging footpaths, I used to note, for my own interest, the patches of snow still clinging on below the highest peaks. October is extremely late for snow to persist from the previous winter in Glen Coe, so I made sure to record the exact dates. I was certain that in the future this would be of interest to someone. How unusually prophetic that turned out to be.

Many years later, in 2005, I read a journal that featured an esteemed Scottish scientist. The paper he authored in the journal focused on the subject of snow patches. By now I was aware of some of the people who engaged in this study, and I wanted to get involved. In the paper the scientist wrote that 'snow has never been known to persist from one winter to the next at Glen Coe'.

Upon reading this statement I cast my mind back twelve years to the diary entry I'd made at Glen Coe. The large patches high on the slopes of the unpronounceable Stob Coire Sgreamhach were visible in October, which is normally a critical month for old snow patches. They are usually terminally small at that point and susceptible to vanishing. There was a very good chance these large patches I saw would have survived through to winter, given the lateness of the year and their size. I couldn't be sure, as I left Glen Coe before the lasting falls of the new winter came. So, with some trepidation, I penned a note to the author of the paper, whose email address appeared at the bottom. The reason for my nervousness was not that I doubted my observations, but because of the reputation and clout of the person who wrote the paper.

Adam Watson, holder of quadruple doctorates and possibly Scotland's pre-eminent outdoors scientist at that time, authored the paper. Here was I: a grunt; an electrician with zero academic pedigree or academic gravitas. I hovered over the 'send' button, about to fire off a note to one of the country's most respected scientists telling him the information contained in his paper might not be 100 per cent accurate. What was I thinking? I shelved the note. But something nagged at me over the next forty-eight hours. I told myself that the diary entry had been correct, and that I'd been looking at these things long enough to know that there was a decent chance those Glen Coe patches survived until winter.

'Send the bloody email, Iain,' I said to myself. So I did, and to my surprise I received a reply the very next day. I sat for a while looking at it unread in my inbox, wondering what it contained. I opened it nervously, through semi-closed eyes. My relief was palpable as I read its contents, which I still have, as they were highly encouraging and warm. A couple of lines stood out:

I am interested to come across a snow enthusiast, as there are so few of us! Best wishes and keep in touch,
Adam Watson

Phew. I was immensely relieved. Here was a scientist who was not precious about the content of his life's work. If data existed that added to the collective knowledge of the study, then he wanted to hear about it, irrespective of whether it meant revisions to his own.

From that point onwards Adam and I corresponded a good deal. We also became friends. Whenever I went out on to the hills, I would seek out places that I knew or thought there would be old snow on, and I shared my observations with Adam. Some of these he would collate and insert into an annual scientific paper of which he was the lead author. The paper, which appeared in the Royal Meteorological Society's *Weather* journal, chronicled how many patches of snow lasted from one winter to the next on Scotland's hills, as well as how long into the year others on lesser hills lasted. This paper was a treasure trove of information to me. In 2008, after I'd been contributing for a few years, I whooped with delight when Adam asked me to become a co-author of the paper. Having my name in a peer-reviewed scientific journal is something that

I never thought I'd achieve. Just a few years later, in 2011, Adam suggested that I have a stab at writing the paper with a view to becoming its lead author. Perhaps sensing that he might not be around for ever, he appeared keen to hand over the authorship to someone who could keep it going for many years to come. I was humbled, and in truth probably more chuffed by this than by anything I'd ever been asked to do before. When my first paper was published, in late 2012, I framed it. A number of years later, in 2017, the Royal Meteorological Society presented us with the annual Gordon Manley Weather Prize for 'outstanding contribution to *Weather* in the preceding five years'. It felt like a real honour to receive recognition from our peers and I attended the ceremony in London with John Pottie, another longstanding expert and co-author who had contributed much to the paper over the years.

Honours like these are very special to us. This is because all the work we do on snow-patch research is done as amateurs. In other words, we do it as a hobby. Every pair of boots, every waterproof jacket, tape measure, rucksack, tank of fuel, even down to our sandwiches – every penny we spend comes from our own pockets. For a living I am employed in construction, specialising in safety and health. I advise the company I work for on what course of action it ought to take to mitigate risk in terms of legislation. It could not be more different from the esoteric world of snow-patch spotting.

Besides, doing it as amateurs has its own advantages. For a start we are not beholden to an institution to spin the results of our research a certain way. In today's cut-throat world of academic funding, leaning towards a favoured position can be crucial to securing financial backing for future studies. Not having to account for our time, or explain to people where we are going, is another distinct advantage. The Scottish weather obeys no calendar. We go to look for snow when conditions are most suitable and, wherever we can, most agreeable. Having to justify ourselves to a supervisor or funding body is something that none of us would feel comfortable with.

Occupying such a small niche does have fascinating consequences. Given the lack of genuine experts in this field, coupled with the increased interest in the subject over the last few years, the level of exposure our work receives has hit unexpected highs. Snow-patch stories have appeared in virtually every national newspaper, as well as on countless radio and TV programmes. This sort of exposure is very helpful for raising the profile of our work, but not just for its own sake. The more people who walk on the hills in summer and autumn, the more eyes there are to report back on the state of any remaining snow. Consequently, the data that we compile and submit to the Royal Meteorological Society becomes that bit more valuable. Without doubt, the number of people who now express an interest in this topic is remarkable. Back in the early and mid-noughties we thought there were only a handful. The upsurge of awareness, not only in enthusiasts but the general public, is most heartening. Long may it continue.

When I first became absorbed in snow patches aged nine, I could have had no idea that it would eventually lead to me being the author of over twenty papers on the subject. It seems to me that you are either born with a love for something or you are not. For sure, interests can be fostered, nurtured and encouraged by parents, guardians or teachers. Generally, though, these tend to be 'normal' or prosaic pastimes like playing an instrument, football or reading. I have yet to come across a parent who actively encourages their offspring to take a tape measure on to the back end of a high cliff or steep gully to measure an eight-month-old patch of snow.

Perhaps that day is coming. Who knows?

two

With friends like these

Some people go to priests; others to poetry; I to my friends.
                                              – Virginia Woolf[1]

The question is *why?* What makes someone become so interested in snow and, more specifically, patches of snow, that they are prepared to go to some of the remotest parts of the country to record relics that are sometimes smaller than the average dinner table? Moreover, what is it that made a young boy become fascinated by the subject?

Childhood interests are often fragile things, being both whimsical and ephemeral. Here today, gone tomorrow curiosities picked up by infants are many but, just occasionally, something sticks. An idea or experience attaches itself to a child, where – allowed to blossom – it can remain for the rest of his or her life. There are an *ad nauseam* number of accounts where respected scientists or naturalists have read or seen something inspirational as children. In their memoirs many of these people seem to be able to recall the event which has gone on to inform their life's study with almost perfect clarity, despite it sometimes being many decades ago.

One beautiful day in August 2012 I set off on a trip to a remote part

---

1   Woolf, Virginia (1931), *The Waves*, London: Hogarth.

of the Central Highlands of Scotland. The ostensible aim of my outing was to check on the status of a large patch of extant snow that I heard lay on one of the area's high hills. When I arrived at the starting point of the walk, some ninety minutes after setting out by bike from my car, I found – somewhat surprisingly, given it was a weekday – a small group of people packing their camping gear away and getting ready for a round of six Munros. They'd come the previous evening, setting up their tents and sleeping under canvas, because they fancied a leisurely start to the day rather than attempting the ten-mile cycle that I had just completed. By the time I'd stowed and locked my bike for the return journey they'd already set off on the path that my route followed. I quickly caught them up and engaged in some conversation typically indulged in by walkers who bump into each other in remote areas. At length one of the walkers noticed the large tape measure strapped to the rear of my rucksack.

'What's the tape measure for?' she asked. I had forgotten it was visible. Normally I kept it in the rucksack so that people didn't ask what I found to be awkward questions. But because I was in the wilds this day, and it was midweek, I hadn't expected to meet anyone. I fumbled around for an adequate answer to the question, but none was forthcoming. It wasn't that I was embarrassed or self-conscious *per se* (though there was a touch of that), but rather that I liked to go about my business quietly, without having to explain my rather individual hobby.

'Oh, just some scientific research,' I replied, hoping that this would deter them from further enquiry. But, of course, it didn't – it only served to make them even more curious.

'What sort of scientific enquiry requires such a large tape measure?' one of the other walkers asked. I knew at this point the game was up and I had better just answer transparently.

'I'm going up to measure the last patch of snow in these hills,' I said. 'It's one of the very few that has been known to last all year outside the Cairngorms or the Nevis range.'

'Oh, how interesting,' another of the walkers said. 'I saw on the news or some other TV programme last year that there were people going out into the hills and doing this. Are you one of these people, then? Well done!'

14

Aside from the slight self-consciousness that I felt (a relic of childhood), the other reason I was reluctant back in 2012 about what I was doing was that I didn't – and, to a degree, don't – fully understand *why* I found it so interesting. For me it is a deeply profound question and one that I have struggled to answer ever since I was a boy. Even now, after all these years of being immersed in the subject and asked about it frequently, I don't think I have ever given a wholly satisfactory or comprehensive reason.

On the specific encounter that day, my main memory is that of relief. The camping party's reaction of interest and congratulations was heartening to hear. That chance meeting on the hill that day was, for me, a watershed moment, and the point at which I realised I needn't worry what people thought, as most of them seemed to be genuinely interested anyway.

Being called a 'geek' used to be a term of derision, attributed to people who were socially awkward or lacking in personable skills. When I was growing up, the word carried baggage that conjured up the image of a lone boy in a bedroom working on a science project or reading books about computer coding. It seems, though, that over the last ten years the word has been largely divested of its negative connotations and is now used far more positively. We geeks have been liberated.

* * *

In the previous chapter I mentioned that the nine-year-old me was sad to see that final fragment of snow disappear on Ben Lomond all those years ago, as though I were waving cheerio to an old friend.

The notion that a particular patch of snow is somehow imbued with a personality, or is in any way anthropomorphised, is patently absurd. And yet, despite this, I have on many occasions set off on a journey to one of my regular long-lying snow haunts, acutely concerned with the condition I would find it in. Why should this be? What does it matter? In the grand scheme of things, it hardly does. It appears, though, that this seemingly odd sentiment is more prevalent than I first imagined.

In his lovely book *Among the Summer Snows*, Christopher Nicholson relays the feeling better than any account I've read. There is a chapter which describes one occasion the author went to the Cairngorms in search of summer snows. A longstanding chionophile, he knew where to look for these patches.[2] But as he stood waiting, the cloud refused to lift, thereby prohibiting a clear view over the valley to the snow. He waited and waited, and then – at last – he was rewarded when the cloud lifted suddenly:

> So here I was, on my knees, late one day in the summer of 2016. My heart in my mouth. Absurd. Is this what it had come to? It: me, the world, human history. The winter would bring more snow, so why did it matter, it did not matter, not really, not at all, the old argument raged away in my head, but at that point it mattered at least enough to keep me there. I needed to know if the snow had survived.[3]

So, why *does* it matter to spotters like Nicholson and me whether snow survives or not? Why do people like us lament its passing and ascribe almost human-like properties to something that is about as far away from being human as it is possible to be? My own view, arrived at after a good deal of reading and listening to psychologists, is that when one develops an interest in something that can border on an obsession, irrational attachments arise. This is especially true for me when visiting the same locations time after time, year after year. One gets to know what a particular patch of snow will look like before one even arrives at it. They tend to take the same shape every year due to the topography of the location they inhabit, so there is a familiarity for the observer. Then there is the fact that the snow is not permanent. It is liable, indeed sometimes likely, to disappear entirely. In this regard the analogy to the human condition is, I believe, apt: the snow is engaged in an existential struggle against time and the elements. It is being chipped away at for every second of every day of its existence until, ultimately, it loses its fight. In other words, from the moment it comes into being, it is always dying.

---

2   *Chionophile* is from the Greek words *chion* meaning 'snow' and *phile* meaning 'lover'.
3   Nicholson, Christopher (2018), *Among the Summer Snows*, Tewkesbury: September Publishing.

To extend the analogy, if one is used to visiting a specific location and the snow is not as large as you expected it to be, then you are invariably disappointed. Sad, even. In the past I have likened this sensation to visiting an elderly relative: you hope that things will be good when you get there, but if the person turns out to be in poorer health than you anticipated, then your own feelings are correspondingly negatively affected.

The patches that Nicholson described earlier as those that caused him to be on his knees are the ones at Garbh Choire Mòr on Braeriach, which I describe fully in chapter 4. I have a longstanding interest in those ones as well, so I can sympathise entirely. But there are others. Ben Nevis's upper reaches possess such drifts, as do its near-neighbours Aonach Mòr and Aonach Beag – both 4,000-feet-plus hills. All these patches are the very longest lasting in Scotland, and the ones that are most visited. Perhaps, therefore, it is no surprise that these are the snows which have been invested with so much emotional capital.

Whatever the reasons are for people attaching themselves to these long-lying wreaths, I for one am glad of them. They help to engender an interest that sees us go back year in, year out. These elderly relatives have family that cares for them.

* * *

There are a multitude of reasons why a hill can cling on to a patch of snow well into summer and sometimes beyond, but it is a multifaceted issue and not always consistent. Even on the same hill, with identical weather patterns occurring over the years, different locations can and often do swap with each other as to which will hold the last drift in any given year. And with temperatures during summer sometimes exceeding 20° Celsius at the locations where the patches sit, people often wonder how *any* snow can possibly resist weather this warm for so long. How can they endure for such protracted periods?

By far the most important reason that affects a snow patch's longevity is the amount of snow that falls in winter and spring at that locality. This may seem like a strikingly obvious thing to say, but it does bear repeating.

If a hollow or gully contains snow that is in the region of thirty feet deep, then it's going to take longer for an 'average' Scottish summer to melt it than if the depth were only fifteen feet. Fifteen feet, or approximately four and half metres if you prefer metric, is in no way unusual for many hills at the end of winter. I have seen depths of old snow in excess of thirty feet in *September* on Ben Nevis's north-east face. Even on relatively modestly sized hills, the right winter conditions can result in extreme depth. Allied to the *amount* of snow that falls, the *direction* that it comes from is hugely important too, as is the *strength* of the wind that propels it. You could quite reasonably say that the quantity of snow is directly dependent on these other two elements.

The second factor that allows snow to persist so long is the physical geography of the hill on which it reposes. In Scotland, as with many other countries in the northern hemisphere, glacial action over the millennia has gouged out corries from the hillsides, many of which face north-east. In form they resemble a large bite being taken out of a cake, especially in the Cairngorms and other areas where rolling hills are prevalent. These bites into the hill are characterised by sheer cliffs and shattered rocks below them. Moisture-laden storms arriving from the west and south-west in winter can, and often do, drop massive amounts of snow above 3,000 feet, with a good deal of it being carried into the corries. This is done, as already mentioned, by the direction and strength of the wind. If there's nothing for the snow to stick to when it falls from the sky, it will be blown across a hill's summit, where it is unrestrained by any obstacles. As it dances along the tops of these hills it is constantly searching for a place to stick and settle. Invariably it will be driven over the edge of the north-east-facing corries and, when it tumbles downwards, it will settle at the foot of the tall cliffs. Each individual storm can add many feet of depth to a single location, and this process repeats numerous times over winter. The very biggest storms in winter can literally add over twenty feet of depth to a single location.

A simple analogy: when one walks in the countryside after a blizzard, fields can appear covered. When, inevitably, the temperature rises by a degree or two during the day, the fields are stripped bare, but the surrounding ditches often contain quite deep drifts. It is exactly the same phenomenon. The fields have nothing on them for the snow to stick to when the wind blows. When it is driven over the edge, it gathers in the ditches and is trapped by the bank. This is why wind is such an important factor in snow's durability. If snow fell vertically on a breathless day it would lie evenly and melt at a uniform rate.

Another reason why north-east-facing patches do so well is that they are, to a greater degree than the rest, protected from the glare of the sun, and the worst of the warm rain, for a good deal of summer. Patches that are tucked in against cliffs that face north-east may only see direct sunlight for the first few hours of the longest days of the summer months. For the rest of the year, the sun traverses the southern sky, its azimuth insufficiently high to reach over the cliffs to trouble the snow. By the same token, the equivalent Atlantic weather systems that deposit so much snow in winter can do the opposite in summer. Southwesterly 'hairdryers', as we call them, from June through to August can wreak enormous damage on the snowpack, but the patches in the north-east corries are protected from the worst of this weather by the cliffs above them.

The third important factor that contributes to snow's longevity is the height the patch sits at. A good rule of thumb is that the higher up something is, the colder the temperature will be. It needs to be said that the higher patches tend to be the ones that survive longest. Generally speaking, if the temperature at sea level is 8° Celsius, it'll be around 0° Celsius at the summit of Ben Nevis. Or, roughly two degrees lower for every 1,000 feet of altitude.

The final aspect that contributes towards a snow patch's durability and is worth discussing at length is a fascinating process called freeze/thaw. It works something like this. When snow falls and settles on the hills in winter, air will be present between the crystals when they land to a greater or lesser degree. Soft, puffy snow is often encountered when temperatures are extremely low and, if further falls of this type occur, then they will simply heap on top of the ones that already exist. While this may make for superb powder skiing and winter pursuits in general, it is of limited value to the snow-patch enthusiast. This is because this type of snow is unconsolidated, or to put it another way, it is fragile and prone to rapid thawing.

For snow to survive long into summer and beyond, a sharp rise in temperature soon after there has been a big deposit of winter snow is desirable. This will have the effect of vanquishing much of the air that is trapped between the individual crystals by the introduction of melt-water and/or rain that invariably accompanies an increase in temperature. What this does is saturate the snowpack, making it heavy and dense. Then, when the air temperature falls to below zero once again (as it is bound to do in winter), the dense, wet pack freezes. This turns fragile snow into resilient snow. Skiers call this kind of snow 'bulletproof' or 'bombproof', and for good reason. When subsequent fresh winter falls cover this 'bulletproof' snow, the process repeats, often many times over the course of a season. The snow will be in a much better condition to resist the summer heat when it arrives as a result of the freeze/thaw process. Consider this example: anyone who has ever tried making a snowball from crystals that have fallen in very cold temperatures will find that they won't bind together very well. No matter how hard you try to pack the snow between your hands, the snowball disintegrates as soon as it's thrown. Make a snowball when the temperature's a bit milder, though, and you'll pack a fair bit of water in with it, binding it together and ensuring maximum effectiveness when throwing it.

Winter 2010 was a very good example of when freeze/thaw played a decisive role in snow endurance. That year, sub-zero temperatures on the Cairngorms persisted for weeks and weeks. Fresh falls of snow were encountered regularly, borne on northerly winds. Soft, puffy snow lay thick and uniform across great swathes of upland Britain. But, as quickly as it came, it vanished when the temperature warmed up. This was purely because the lying snow was unconsolidated. Between the falls of fresh snow there had been very few – if any – rises in temperature, so the pack remained fragile. Warm weather removed it rapidly and what might have been a vintage year for snow-patch enthusiasts turned into a poor one, with very few patches surviving the whole year.

What of the depths, though? Just how deep can snow get in Scotland? As alluded to previously, if the right conditions are met then accumulations that most would find almost unbelievable can build up. The very deepest that I've seen in this country exist, unsurprisingly, on Britain's tallest hill, Ben Nevis. It has certain attributes that confer on it advantages other hills cannot compete with. Firstly, and obviously, is its height of 4,411 feet. Where rain or sleet falls on a hill 1,000 feet lower, most of the precipitation that lands on Ben Nevis between November and April will be snow. Secondly, its proximity to the west coast of the country ensures that it gets both barrels of the Atlantic storms that can drop thousands of tonnes of water or snow every hour. Thirdly, the plateau of the hill allows for the phenomenon described earlier, whereby snow races off the summit and on to the cliffs and gullies below. And what gullies below! Dozens of them snake from the summit plateau, running sometimes for over a thousand feet. The biggest of them all, Observatory Gully, carves a path from just below the summit cairn all the way down to the floor of Coire Leis, 2,000 feet lower. During the very snowiest of years, depths of 100 feet would not be surprising in the heart of Observatory Gully. For context, 100 feet is the equivalent of seven double-decker buses stacked on top of one another. 1951 and 1994 are two years when accumulations may have reached this remarkable total.

* * *

What is it that I and others *do* when we get to a patch of snow in summer and autumn? What we are primarily concerned with is what size the patch is and, more importantly, what date it is likely to vanish, if at all. If the snow has already gone when we arrive, then an estimate is made as to when we think it disappeared. For the former, there is no hard and fast rule that can be applied, and nor is there a formula. Calculating how long a patch will last is based on its size at the time of the visit, what the weather does afterwards, and applying the human factor of judgment based on experience. Concerning the latter, it is unquestionably disappointing to trudge for miles and climb to 3,500 feet, only to find no snow. However, it is just as important to prove a negative as a positive, so no trip is wasted effort. Often the vegetation surrounding a recently deceased patch can give clues as to when it did finally disappear (flat and dormant grass, waterlogged and semi-frozen terrain). It is important to log the final melt date of a patch so long-term data can be built up. Recording the same one for, say, twenty years allows a broader perspective on any trends, should they exist.

In terms of actual fieldwork and planning which areas to visit, this is now very much a product of experience. Though it is always useful to keep an eye on how much snow falls over winter, my work generally doesn't start in earnest until April. One gets to know certain areas where snow vanishes relatively early (April or May), so attention can be focused there initially. It is possible every year to find snow in April on even modestly sized hills in northern England and southern Scotland. In exceptional years, such as 2013, relics of winter can be found in May on the relatively low hills of the Peak District. The chionophile must always be ready to go somewhere in the briefest of weather windows, or if an unlikely patch has been spotted in a curious location. More recently, social media has proved enormously helpful in this regard. There are now scores of people who actively keep tabs on certain hills, sending me pictures via Facebook or Twitter. Even people who post pictures of hill walks on internet forums (such as *www.walkhighlands.co.uk*) often unwittingly provide invaluable data when the summit photo they have taken shows a patch on some distant hill, over the shoulder of the grinning walker-cum-model. All this allows forward planning, with visits far less haphazard than they used to be. Even fifteen years ago, when I started assisting with Adam Watson's annual surveys, we'd go to a hill not even having a clear idea if we'd find any snow. This may have been part of the charm, but as one grows older, charm is gratefully sacrificed for accuracy.

In the last few years satellite imagery, too, has appeared and now plays a decisive role in deciding where to go. Providing it's clear, which is annoyingly rare in Scotland, the Sentinel satellites give a reasonably good perspective, and can identify individual patches of snow down to about ten metres in length. Without doubt, technology has made the art of snow-patch observation more precise. It is rare now that I go somewhere and am genuinely shocked that the snow has gone or is far larger than I'd anticipated.

Knowing where on a hill to look for the elusive relics of winter, however, is part of the fun. Most of the time this isn't as difficult as it might sound. Because of a hill's topography, patches virtually always form at the same place and take almost exactly the same shape. There are minor variations, of course, but year after year it never ceases to amaze me how uniform their outlines are. But it is not just deciding where and when to go that has been made easier by technology. In terms of the equipment required, no longer do I have to carry a 100-metre tape measure on the outside of my rucksack. These days laser tape measures, which weigh a fraction of the manual equivalent, do a better job. It is also nice not to have to fend off questions from people who wonder why on earth my rucksack is so light that I feel the need to strap a large tape measure to it. (And I will not miss the innumerable times that I weighed down my tape with a large rock at one end of an eighty-metre snow patch and walked across precarious ground using one hand for balance while holding on to the body of the tape measure in the other, only for the tape to ping off just as I was about to reach the other end of the patch.) Predictably, the ubiquitous mobile phone has removed the need for carrying a host of other heavy gear, including notepads, Dictaphones and, to a lesser degree, cameras – though I never go to the hill without a proper camera. As good as modern phones are, the depth of field and lens quality means they are far inferior to a proper camera.

Generally speaking, snow recedes northwards. Therefore, eyes are trained in spring on the Southern Uplands (all the high hills south of Scotland's Central Belt).[4] No hill in this range exceeds 3,000 feet, though, and this combined with their rounded summits and lack of proper cliffs means that winter's snows are quickly diminished in the spring. There is little shelter from sun, wind and rain on offer here. A rare exception to

---

4   I also write a paper each year for England and Wales.

this rule is found on the second-highest hill in the Southern Uplands, Broad Law. On its north-east face sits a rough, rocky escarpment called Polmood Craig. Snow here occasionally reaches proportions that can rival higher hills much farther north in the Highlands.

Once the Southern Uplands' last snows have expired, it is time to focus on the Highlands. This is when it becomes altogether more difficult. In the hills of southern Scotland, the same four or five vie with each other to see which can hold on to the last snow. North of the Highland Fault Line, all bets are off.[5] Eyes are needed absolutely every-where, so trying to decide where to visit takes a good deal of research. Such is the complexity of the terrain, and such are the sheer *numbers* of locations that can hold snow, that there needs to be an element of circumspection. No person can hope to cover all areas, even if they were employed full-time to do so. It may be possible to do that later in the year, when the number of snow patches is fewer than fifty, but from April to August it is simply not realistic.

In an ideal world I would be in charge of four or five full-time observers who were on the payroll, each keeping tabs on his or her designated area. This would ensure that far more data could be collected and compiled. The data would, ideally, then be analysed by the best minds in the country, the better to build up a wider resource base. Short of a huge lottery win, the chances of this happening are almost zero, but I can dream.

---

5   The Highland Fault Line is a geological fault line stretching across Scotland in a north-east direction, separating the Highlands from the Lowlands.

three

The footsteps of others

If sport like this can on the mountains be,
Where Phoebus flames can never melt the snow;
Then let who list delight in vales below,
Sky-kissing mountains pleasure are for me.

– John Taylor[1]

… the upper and innermost part of the country, where the tops,
or summits of the hills are continually covered with snow, and perhaps
have been so for many ages, so that here if in any place of the world they
may justly add to the description of their country:
Vast wat'ry lakes, which spread below,
And mountains covered with eternal snow.

– Daniel Defoe[2]

1   Taylor, John (1618), *The pennyles pilgrimage, or the money-lesse perambulation, of John Taylor, alias the Kings Majesties water-poet*, London: Printed by Edw. Allde, at the charges of the author.
2   Defoe, Daniel (1726), *A tour thro' the whole island of Great Britain, divided into circuits or journeys [etc.]*, London: D. Browne [etc.].

Historical sources for Scottish snow observations, before the era of Victorian rigour, tend to be infuriating and tantalising in equal measure (if no less entertaining for all that). When I was engaged in research for a book I co-authored back in 2010, I used to frequent the reading rooms of the British Library in London. The number of instances where I audibly cursed a writer for their slipshod reportage are too many to remember. There I'd be, on the edge of my seat, reading a wonderful eighteenth-century account of a writer seeing snow on the tops or cliffs of some hill or other, only for them either to omit the day or – unforgivably – the month that they observed it.

The two opening quotations are both classic examples of this. In the first, written during the early seventeenth century, John Taylor wrote about 'Phoebus flames' being unable to melt the snow (i.e. the snow was perennial) that lay on the hills in the locality of Braemar – which is where he was hunting with Lord Erskine when he penned the short poem.[3] It is too easy to dismiss Taylor's flowery description of the snow being permanent as mere romantic embellishment. I'm not at all sure it was.

When Taylor wrote this verse, the country was well in the grip of the Little Ice Age. It is almost certain that many of the high hills around Braemar – the southern Cairngorms – would have carried far more snow in summer than they do now. Maybe he was merely reporting the hills as he saw them, or even recounting a story told to him by his host Lord Erskine, which prophesied the peer's tenure of land being lost if all the snow were to vanish. Echoes of this legend can be heard right to this day.

The major landowners in the Braemar area are now the Farquharsons of Invercauld. One of the highest hills on Farquharson's land is Beinn a' Bhuird. On it, facing south, sits a highly conspicuous snow patch, called for many years by indigenous residents 'the Laird's Tablecloth'. According to local legend, the Farquharson family will lose their lands of Invercauld if the snow ever vanishes. The most recent laird, Captain Alwyne A.C. Farquharson, said during a radio interview in 1982 that:

---

3   Phoebus is another name for the Greek deity Apollo.

so long as the Tablecloth ... remains spread, the Farquharsons would hold their lands. And that's the reason why no Farquharson will ever admit that the Tablecloth is not spread. They might go so far as to say that the linen was dirty and needed a wash, but if you take the trouble to climb up and have a look you'll find it spread, summer just the same as winter.[4]

The last sentence is clearly tongue in cheek, because the Laird's Tablecloth almost always vanishes nowadays. But for such a legend to gain a foothold in the minds of Deeside locals there needed to be a basis in fact. So, if we take Taylor at his word, we could easily believe that he was not indulging in a bout of overexaggeration by asserting that 'Phoebus flames can never melt the snow', but, rather, he was reporting on how it was at that time.

Equally, in the second account, Defoe's story of hills' summits being 'continually covered with snow' seems to have a whiff of truth about it. The book in which he describes the countryside is not filled with hyperbole, so it is a reasonable inference that what he portrays is accurate. Having said that, it is unlikely that the summits *per se* would have been 'continually covered'. It is more plausible that it was the cliffs and hollows around them. As outlined in the last chapter, snow is far more attracted to these features than it is to summits.

* * *

There are different types of author that an interested researcher can draw on when he or she digs into the archives of historical snow literature. The first is the one exemplified by Taylor and Defoe: people who compose quite lovely verse but are next to useless at providing reliable dates. (In Taylor's case he didn't even give the *year* he travelled.) Another is the eighteenth- and nineteenth-century travel writer. These authors tend to be far more reliable in what they are reporting on. True to the spirit of the age, where adventure, shooting and having servants were highly fashionable, there is a relative abundance of these.

---

4   The interview was conducted by Adam Watson and passed to the author.

Many writers of that age ventured to the wilds of Scotland from their leafy, west-London villas in search of romantic and heroic characters. Often they'd return south disappointed, having failed to find what they were looking for. Several more complained about the weather and uncouth locals. (At the time many indigenous Highlanders lived in very poor accommodation, sharing their houses with not only a pig or a cow, but associated vermin.) Of the ones that toughed it out, however, their accounts can be fascinating, and not just for the references to snow. Usually there are wonderful cultural asides thrown in, which give a unique window into how everyday folk lived.

Among the most informative of the travel writers of this era was a quite remarkable woman called Mrs Sarah Murray. Having married into upper middle class London society, she was clearly not short of funds nor inclination. This is amply testified to by her ability to spend a start-ling total of five months travelling around the UK with two servants. Clearly very well educated, she provides vivid descriptions in her book of what she found. One of the best of these recounts a trip she made to Cairn Gorm in 1801. After taking a pony up to the summit of the hill on 6 September, she observed that:

Those who mount that eminence [Cairn Gorm] should walk to the edge of the precipices hanging over the hollow towards Loch Auon [Avon] and to the snow house. The snow house is not far from the Cairn or heap of stones, on the highest part of Cairngorm, and is a hollow, in extent an acre or two. This hollow is filled with snow, and although it faces the south, it is never melted either by the sun or rain ... in its bed are large stones standing high and thick, serving for supporters to the roof of snow, which seems to be in some degree petrified ... I therefore, by bending my body, walked up the bed of [the] rivulet for three or four yards ... but it was so intensely cold under snow and in water, that I was obliged quickly to return.[5]

5   Murray, Sarah (1803), *A Companion and Useful Guide to the Beauties of Scotland [etc.]*, London: printed for the author.

This superb snippet is of far more value to researchers than those of Taylor and Defoe. Not only for the detailed description of where it is, but for the accurate date and what was encountered when she visited. Reading it over 200 years later, several things jump out. The location of her trip is, almost certainly, a place that is now called Ciste Mhearad (Margaret's Coffin). It is no longer called 'the snow house', if it ever were called that. This sobriquet may have been a name invented by Murray herself. Ciste Mhearad is situated precisely where she portrays it, close to the summit of Cairn Gorm, the hill that gives its name to the whole range of the Cairngorms. She says it (the 'snow house') faces south, but it is more east than south. This is but a minor quibble. 'It is never melted by sun or rain' is clearly a reference to the snow being permanent. Murray wouldn't have known this, of course, as she was just a visitor. It may be assumed with certainty that she had a guide on this trip, one who may have relayed the information on the snow being permanent. Indeed, for her to venture there in the first place suggests that the local folk knew this location well, and its proclivity to hold snow all-year round. It is invisible from the Spey side of the Cairngorms, which would have been the largest population centre at the time. Clearly, this was a place unusual enough to warrant guides taking people to it on horseback.

The 'rivulet' she walked up unquestionably refers to a snow tunnel of the type commonly found in Scotland in summer and autumn, where stream courses gouge out lying snow, sometimes to a height exceeding fifteen feet. We may be certain that Murray's story is not confected because of the accuracy of her portrayal of this feature. She could not have described it so well without direct experience. For such a tunnel to form at this location in September nowadays would be unthinkable, and it is evident that far less snow is present here nowadays than there was then. Rather than being permanent, today it is only a very occasional survivor. Consequently, Murray's account is of incalculable worth to researchers of weather phenomena. But that was not all Murray had to say; remarkably, this was her second visit to the locality in the space of several years. Indeed, it is likely that her journey to 'the snow house' was inspired by her first visit, where, a few years previously, she noted:

the house [Rothiemurchus] is within sight of Cairngouram [Cairngorm] mountains, whose hollow cliffs are filled with never-melting snow. The cap of winter upon the crown of luxuriant smiling summer below, was a contrast I had never before beheld ... The sun, however, was sufficiently high to gild the mountains and the lovely scenes around Rothamurchus; and for many a mile my eyes were feasted by the white patched hollow sides of Cairngouram.[6]

The view she describes, near to where the current town of Aviemore is situated, is on an early summer's day almost unsurpassable for beauty from an urban environment in the UK. High native fir trees stud the landscape, and Braeriach's northern corries' cliffs grab the viewer's attention especially. They are normally flecked with snow until well into summer and are perfectly described by Murray. It is little wonder she chose to return a few years thereafter.

Another traveller of this period was a redoubtable and eccentric character by the name of Colonel Thomas Thornton. In an extraordinary essay from 1804, *A sporting tour through the northern parts of England and great part of the Highlands of Scotland*, which chronicles him, frankly, rampaging around the Highlands on a prolonged shooting frenzy, Thornton depicts his travels from previous years. One such trip recounted an unconventional journey up on to Sgòr Gaoith, a high hill in the west of the Cairngorms. The book itself is laced with self-importance and bravado (and a *lot* of shooting), as befits an eighteenth-century former army officer. As a result of this bravado, one must exercise an element of circumspection when picking out what is factual. The account in question, though, would appear to be broadly reliable, based on the landscape that he describes. In 1786, Thornton and his party travelled on 6 August from Ruthven near Kingussie to cross the River Feshie and climb to the ridge of Sgòr Gaoith above Loch Einich. In the morning they agreed to make 'one of the gullies of snow' (i.e. there was more than one gully with snow in it) their guide. This would have been a snow

---

6   Ibid.

patch on the Glen Feshie side of the ridge. The party proceeded on a very hot day, when Thornton recorded 29° Celsius in the glen, and at twelve o'clock they arrived at the first snow. On reaching the high ridge, they were:

> depositing our champaign, lime, shrub, porter, etc. in one of the large snow-drifts, beneath an arch, from which ran a charming spring.[7]

This is doubtless Fuaran Diotach or the 'dinner well' south of Sgòr Gaoith. Again, it is revealing that he wrote 'one of' the large snowdrifts, plural. He described the higher mountain above them (Braeriach) as 'chequered with drifts of snow'.

Accounts like Thornton's and Murray's are interesting because of the *by the way* observations they make about snow. To visit these old drifts is not their objective in the wider context of their tours, so it may be assumed that there is no compulsion for them to embellish the story. In August nowadays, the location where Thornton and his party deposited their champagne would have long before lost its snow. To see these patches, as he did, on the Glen Feshie (i.e. the west) side of Sgòr Gaoith is unheard of in this era. It is yet another piece of, albeit anecdotal, evidence that points towards the current epoch not getting as much snow in winter as during the Little Ice Age, and our climate being quite different. Another entry from Thornton's journal from 9 September 1786 is eye-opening:

> Began to shoot at eleven, but found it wonderfully cold, when, looking over towards the Cairngorms, I discovered a very sufficient reason for the chilliness I experienced; for the wind was not only north, but the heights were completely covered with hillocks of snow, which appeared, as well as I could judge from their distance, to be at least four feet thick: we had some sprinkling of snow before; but this, I confess, astonished me.[8]

7 Thornton, Thomas, *A sporting tour through the northern parts of England, and great part of the Highlands of Scotland* (1804), London: Vernor and Hood [etc.].
8 Ibid.

In living memory such a depth of snow falling in early September has happened only once, in 1976. It may well be the case that Thornton's experience was similarly rare, but it seems unlikely.

* * *

Some of the other snow-related anecdotes from the period of 1750–1900 also provide fascinating insights into the cultural practices of the day. Parish minister the Reverend Donald McGillivray wrote in 1835 about Ben Nevis in the highly informative *New Statistical Account*:

> The deep clefts on the north-east side of Ben Nevis are never without snow. For two seasons when ice failed, the snow gathered and condensed into ice in these clefts and was of great service to the salmon-curers. The country peasants with their small hardy horses carried it down in panniers on horseback.[9]

This very much understated comment is worthy of elucidation. For fishermen to even *consider* going to the cliffs of Ben Nevis to gather snow shows how important a part it played in the workings of the local economy. Anyone who has ever been to the snows on the north-east face of Ben Nevis knows what is involved. They are situated above testing terrain, characterised by greasy and unstable rock which is liable to give way at any time. On foot the snows are hard to reach, even today where a good footpath has been constructed between sea level and the refuge at 2,200 feet. In the early nineteenth century, taking ponies up rough ground to above 3,000 feet must have been a significant and risky undertaking, necessitating the best part of the day to get up there and back. As an aside, it would be fascinating to know just how many ponies they took up, and how much snow they brought back in their panniers. Clearly it was a trip worth doing, so one would imagine it was quite a few ponies.

One of my very favourite historical accounts of a colder Scottish climate was written by 'that Intelligent Knight Sir George Mackenzy' in 1675:

---

9   *The New Statistical Account of Scotland* (1834–1845), Committee.

There is another little Lake in Straglash at Glencannich on lands belonging to one Chissolm; the Lake lies in a bottom 'twixt the tops of a very high hill, so that the bottom itself is very high. This Lake never wants [i.e. lacks] Ice on it in the middle, even in the hottest summer, though it thaws near the edges; And this Ice is found on it, though the sun by reason of the reflexion from the hills in that country, is very hot, and Lakes lying as high in the neighbourhood have no such Phenomenon. 'Tis observable also, that about the borders of this Lake the Grass keeps a continual verdure, as if it were in a constant spring and feeds and fattens beasts more in a week than any other grass doth in a fortnight. The matter of fact I have fully examined in both these; but to hit the cause, requires a better Philosopher than I am.[10]

The 'little Lake' being referred to is almost certainly Loch Uaine ('green loch') between the two highest hills in the north-west Highlands, Càrn Eighe and Màm Sodhail. Sitting at an altitude of just under 3,000 feet, it is one of the highest bodies of water in that area. It held broken ice at the start of June during the 1900s, but only after very cold snowy winters and springs, notably in 1951 and 1977. No one currently alive has seen ice persist there till July.

As for the 'feeds and fattens beasts' comment, this adds independent weight to the credibility of his observations about ice. What he came across was the fact, now well-known from studies on arctic tundra and alpine land, that vegetation which has recently emerged from snow or ice quickly shows a flush of spring growth, even though the time may be late summer or autumn. On the Cairn Gorm plateau, for instance, studies have found that grazing sheep strongly favoured grass near the edges of long-lying snowbeds. Within three metres of the edge, soil was frozen or waterlogged, and plants dormant, whereas favoured plants at three to ten metres had started to grow. Plants further out than ten metres had grown for a few weeks and had passed the first growth flush, and sheep

---

10 Macfarlane, Walter (1907), *Geographical collections relating to Scotland made by Walter Macfarlane*, edited by A. Mitchell & J.T. Clark, Edinburgh: Scottish History Society.

avoided them. So, as snow receded, grazing sheep ate newly growing plants, rather like a long spring season. Sir George's 'constant spring' was an apt observant phrase more than three centuries ago.

* * *

Of the quirkier tales of historical demands that recur in Scotland, none is more interesting – or charming – than the monarch insisting on snow at midsummer. Indeed, in a tradition similar to the Farquharsons of Invercauld – who, it may be recalled, owned their land on the proviso that the Laird's Tablecloth was set – a number of other families' landed interests apparently depended on snow. The tradition appears to be extremely deeply rooted in Highland folklore, because it occurs widely. For instance, the Comyn (or Cumming) family of Lochaber was given the Lordships of Lochaber and Badenoch about 1229 and their 'right' to the Lochaber lands was as long as there was 'Snow on Ben Nevis, heather on Druim Fada, and ebb and flow of Loch Eil'.[11] This tradition endured for hundreds of years and was noted again in the early nineteenth century: 'It is said that Cameron of Glen Nevis holds his lands by the tenure of an unfailing snowball when demanded.'[12] The story was fleshed out in a book from 1986 on the history of Ben Nevis. It stated that:

> an old legend, shared with some other Scottish mountains, has it that should all the snow vanish from Ben Nevis then the land shall be forfeit. To prevent this, Cameron of Locheil, in a summer when it looked as if the last of the snow might melt away, is said to have sent locals up the mountain with straw, to protect the remaining snow patches.[13]

The Munros owned the east side of Ben Wyvis, near Inverness, and an account states that for centuries they held their land on condition that

---

11   MacCulloch, Donald B. (1971), *Derivation of the Name Lochaber*, Clan Cameron Archives.
12   MacCulloch, John (1824), *The Highlands and Western Isles of Scotland*, Edinburgh: Longman [etc.].
13   Crocket, Ken (1986), *Ben Nevis*, Glasgow: The Scottish Mountaineering Trust.

they 'supplied a bucket of snow at the Palace of Holyroodhouse on Midsummer Day to cool the King's wine'.[14] (How the Munros managed to get the snow from Inverness-shire to Edinburgh before it melted is not explained.) In the same book it was said that the Macintyres of Glen Noe in Argyll paid their rent, on Midsummer's Day, of a 'white-fatted calf, and a bucket of snow' taken from a high corrie of Ben Cruachan. Lastly, a book from 1951 gives a further three examples: 'Certain Highland families traditionally hold their land on the condition that they are able to supply the King with a bucket of snow whenever he should pass that way. The Camerons of Glen Nevis, the Grants of Rothiemurchus, and the Munros of Foulis are all supposed to hold their lands on this condition.'[15]

Though in the main anecdotal, these stories – for their independence of authorship – are unlikely all to be confected. Some may contain, to a greater or lesser degree, an element of exaggeration, but it seems a strand of truth weaves through the stitching of these legends. The stories are consistent with what seems to be an unarguable fact: that our winters are far less snowy than they used to be. Or, to summarise *Cool Britannia*, the 2010 book I co-authored with Adam Watson:

> Although much of the above historical evidence is anecdotal and qualitative, it appeared independently with many observers in different years and regions. Because of this, we infer that observations were unlikely to be atypical, such as being concentrated by chance in an unusually snowy year or area. Comparisons involving quantitative observations from recent decades add confidence to this inference. To sum up, it is reasonable to conclude that more snow lay on the hills of Britain during the 1700s to early 1900s than in decades since 1930. Accounts from the lowlands of Scotland and England fit with this.[16]

* * *

---

14 Gordon, Seton (1971), *Highland Summer*, Abingdon: Littlehampton Book Services Ltd.
15 Gordon, Seton (1951), *Highlands of Scotland*, London: R. Hale.
16 Watson, Adam, and Cameron, Iain (2010), *Cool Britannia*, Bath: Paragon Publishing.

As interesting as these older historical references are, and as much as I admire the fact that they bothered to chronicle them at all, later commentators compiled their records rather more assiduously. These are the people who were the true pioneers of the study that I and others are now engaged in. They were diligent and thorough, keeping superb records. The work that I, and people like me, do tries to follow very much in their footsteps.

Chief amongst these pioneers are two remarkable men, Seton Gordon and Gordon Manley. The former is well known in Scottish outdoor circles. Born in Aberdeen in 1886, he was a superb naturalist and writer. His inherent enthusiasm for wildlife and writing was no doubt improved upon by the private education he enjoyed, and his time at Oxford University. Seton Gordon knew many of the landed families in Scotland and moved in their circles. He had a wide variety of passions, one of them being long-lying snow patches. Luckily, he left a great store of highly valuable information, wonderfully written. He was active at a time when comparatively few people took to the hills, so his observations are of great worth.

Gordon Manley is perhaps less well known than Seton Gordon. However, his body of work on the study of long-lying snow is almost without equal. A native of the Isle of Man, Manley was born in 1902 and grew up in Blackburn, Lancashire. He read Geography at Cambridge University and set up many meteorological stations across the country. He was a founder member of the Association for the Study of Snow and Ice which established the National Snow Survey. His 'The Snowline in Britain' paper, which appeared in a Swedish journal in 1949, is a *tour de force* of its genre.[17] In it he not only discusses snow patches on Ben Nevis and elsewhere, but also how much of a drop in temperature would be required for glaciers to reform on some British hills. From his many books and scientific publications, as well as his field observations, I feel sure that Manley and I would have got on like the proverbial house on fire.

Seton Gordon seems to have been a more complicated character.

---

17   Manley, Gordon (1949), 'The Snowline in Britain', *Geografiska Annaler* Volume 31, 179–193.

Adam Watson told me much about him, as they met and corresponded frequently from the 1930s right up until Gordon's death in 1977. Despite his complexities, Gordon's writings on the general outdoors – and on snow – were in a league of their own. Nan Shepherd is often lauded as the best writer on the Cairngorms, but for me Gordon's prose captures the essence of these hills more satisfactorily. A snippet of his work, taken from a night-time sortie to a remote part of the Cairngorms, shows this well:

There was silence in the Lairig, save for the croaking of some ptarmigan and the rush of the March Burn. Exactly at midnight we reached the watershed and the Pools of Dee. For some time, the sky southward had foretold of the rising of the moon, and now, at midnight, the pale orb climbed above the shoulder of Ben Mac Dhui and shone into the dark pools that lay as though asleep ...

In early summer the snows of the Garrachory are as yet almost unbroken, and where the moon shone the snows were faintly tinged with gold ...

We reached the head of the corrie, where snow lies from one year's end to another, as the first pale flush of dawn appeared in the sky to the north-east. Behind the black cliffs to the west the cirrus clouds were tinged with lemon where the invisible moon shone, but as we climbed we at last saw the moon herself (dimming her lamp against the dawn) and sat among the confused waste of immense granite boulders to await daylight.

All around us in the half-light were great snowfields, pale grey and mysterious. Up the rocky slopes of Sgor an Lochain Uaine a herd of phantom deer seemed to strive upward, yet ever to remain in the same place. To the unaided eye it appeared that they swayed this way and that in confusion, but the glass showed the black army to be rocks projecting from the snowy surface. From the western sky two stars shone palely, their rays becoming momentarily more dim as the daylight increased.[18]

---

18   Gordon, Seton, 'The Garrachory by Night', *The Scotsman*, 3 July 1926.

After the 1950s, the number of people interested in observing long-lying snows increased. In part this may have been to do with increased leisure time and opportunity of travel, and no doubt greater disposable income some had during the post-war construction efforts. Because of this, and as the people engaged in it generally did so with a bit more precision, the corresponding records are more valuable. Some observations are better than others, however.

One well-known spotter of the era was a gentleman by the name P.C. Spink. He wrote a paper annually from 1965 to 1980 for the Royal Meteorological Society, chronicling the 'Scottish Snowbeds in Summer'.[19] I have a great deal of respect for Spink and his obvious enthusiasm. Unfortunately, his methods and – therefore – his conclusions, were rather dubious. He lived in Lincolnshire and came up to Scotland only infrequently. Sometimes his pronouncements of which patches of snow had survived the year were based on observations made in July or August. These are months too early to be considered reliable. He often over-estimated or simply missed patches because he wasn't observing from the right place. For example, in August 1964 from Ben Macdui he saw 'no signs of snow survivals on Ben Nevis' but most of the Ben Nevis patches that last till winter are invisible from Ben Macdui. Therefore, despite the relatively large body of work Spink produced, it would be sensible only for information on those sites that he visited at close range or in good visibility on the stated dates to be considered as authoritative.

Two observers from recent decades, John Pottie and David Duncan, were local to the high hills which hold the longest snows. Their homes afforded good views towards these hills, and because they were able to go to them frequently and at short notice, the quality of their records is generally superb. They both took meticulous notes and photographs, especially of the Cairngorms. Our knowledge of survivals from year to year is greatly enhanced by their diligence. For many years both were co-authors of the annual snow-patch paper that I now write. Pottie camped in the snow at 4,000 feet when he was in his seventies, just so he could

---

19   Spink, P.C. (annually from 1965 to 1980) 'Scottish Snowbeds in Summer', *Weather*.

get a sighting of an old patch before it was buried for the winter season. Such dedication is rare even in people less than half his age.

One man, though, has done more for snow-patch observations than any other. Though he has been mentioned already in the first chapter as the person I contacted all those years ago, Dr Adam Watson's work deserves a whole book on its own. I devote an entire chapter to some of it later in the book, especially his observations on snow. He was a visionary and all future work on this subject will continue along the tracks that he laid almost single-handedly.

* * *

That so many historical writers even bothered to mention seeing snow on the hills suggests that the British interest in it is deep. During the writing of our 2010 book, Adam and I came across scores and scores of historical references to snow when travellers were sojourning around the Highlands. Often the references were in passing and of little use as factual observations. Others, as we have seen, were superb.

It is perhaps our latitude and proximity to the Atlantic that fuelled our interest in snow during the past and continues to do so to this day. In many countries, such as Norway or Sweden or Canada, snow is more or less a banker. It comes regularly, so is expected. High hills in those countries carry glaciers, which mean that individual patches that persist are unlikely to attract much attention. (I remember when I went to Egypt in 2001, I woke up on the first day and said to the guide, 'What a lovely day.' He simply shrugged and said that every day was like that and no one talked about it.)

In the UK, however, snow is an occasional visitor, rare enough to be a surprise. It is also hard to predict. It might arrive at the end of October, even at relatively low levels, but it also might not be seen at all until late March, if even then. There is simply no rhyme nor reason to when it shows up. Maybe because of this we find it intriguing, and that is why it has attracted a level of interest from writers over the years out of proportion to its scale.

The final factor that ought to be considered is the slightly eccentric nature of Brits and British hobbies. There is definitely a touch of Noël Coward's 'Mad Dogs and Englishmen' about this pastime. Yes, I am Scottish, and so are many of the people who indulge in this hobby. But I recognise something of that mad-dog tendency that encompasses not just Englishmen but Scots, too. A story I am fond of retelling, which highlights this odd national trait, happened in 2001. In November of that year, twelve British tourists were arrested by Greek authorities after being accused of spying near an airport in Kalamata. The reason given to the Greek authorities by the tourists and the British Embassy was that they were on a plane-spotting holiday. As a pastime, plane-spotting is unheard of in Greece. Their excuse was met with contempt. The police were simply unwilling, or unable, to believe that a group of twelve people would fly from Britain to Greece just to stand at the end of a runway with binoculars and notepads. Surely nobody would be so, frankly, weird. The whole thing was put down to a massive cultural misunderstanding.

I am unlikely ever to be incarcerated for my research, but I do so empathise with my plane-loving countrymen who were jailed for their passion.

four

Braeriach:
place of the eternal snows

The bulk of Braeriach is immense. Apart from Ben Nevis, its only possible rival, the Cairngorms' second-highest hill has no British equal in terms of complexity or majesty. Its name, translated into English as 'the brindled upland', understates its character significantly. There may be other hills in Scotland that are, arguably, more photogenic upon first glance: An Teallach, Liathach, Suilven, Ladhar Bheinn, Bidean nam Bian. All these peaks, with their airy ridges and pointed summits which tear holes in the sky, take the breath away. From many angles Braeriach lacks the steep upwards projection that draws so many photographers to the West Coast in search of the perfect calendar shot. It has a languid, torpid majesty which contrasts starkly to those farther west, ones that seem to huddle close to their neighbours as though fearing their own solitude. Braeriach's various minor tops lean back and stretch, rather than stand upright like a phalanx of granite sentinels.

From no matter which angle one gazes upon the slopes of Braeriach, the eye can never settle. It is drawn here and there by a cliff or gully, lochan or snowbed. Its summit, some 4,251 feet above the sea, lies just a few yards from the shattered granite cliffs of Coire Bhrochain, which themselves fall away for over 1,000 feet. Depending on how one counts them, Braeriach has another twelve corries to go with Bhrochain.

No single British hill, not even the huge primeval edifice of the nearby Beinn a' Bhuird, can compare.

Each one of these thirteen glacial gouges in the hill possesses its own character. Though two may sit adjacent to each other, separated only by a wall of granite, they will be utterly different. Take, by way of example, the magnificent Coire an Lochain. Walk in April by the loch that sits in its bosom and you will be transported to within the Arctic Circle in early summer. In this corrie, winter does not give up easily. As spring's warmth heats the water of the loch, the ice sheet – which covers it for at least five months of the year – starts to reduce in mass. Large ice floes form as the sheet breaks up, and it takes only a little mental embellishment to imagine a hungry polar bear standing atop them, licking its lips as it waits for an unsuspecting seal to surface for air. The cliffs above the loch hold snow late into the year, easily visible from the town of Aviemore, miles to the north.

Coire an Lochain has two neighbours to the east: Coire Ruadh and Coire Beanaidh. Though both, maybe, lack the grandeur and beauty of Coire an Lochain, the streams that issue from them combine to form a watercourse, the Beannaidh Beag, that can be as fierce and beautiful a barrier to a snow-patch observer as any high cliff. In full spate, just as it enters the larger Am Beannaidh river, the ford across it is insuperable. I recall some years ago having to adjust my plans of a sortie to the long-lying snows of Braeriach on account of the foaming torrent being a one-way ticket to a certain early grave. Granite boulders weighing a third of a tonne had been gouged from the riverbed by the storm and tossed to the side. Had I gone there just a few days earlier I'd have been able to ford the river without bothering to remove my boots. Such are the vagaries of the Cairngorms' weather systems.

High above the Beannaidh Beag's ford, Braeriach's huge and open plateau stretches for miles. It is not so much a typical Scottish summit but, rather, an Arctic-like desert where only the hardiest plants can get a toehold and the doughtiest animals inhabit. Loose gravel, broken down by many thousands of years of weathering, lies haphazardly among the larger rocks that nature has not yet crushed. Skin-flailing winds are commonplace. But, for all its bare harshness, the plateau has its own beauty.

On a March day a few years ago, I zigzagged on skis 2,000 feet up a gully from Glen Eanaich on to the plateau, only to be met by thick and heavy cloud. Snow cover on the plateau ran complete, but flat light made the going difficult and hard to read. Care was required near the eastern rim, where multiple corries yawned and bit deeply into the hill. On terrain such as this many people have paid for misnavigation with their lives. When one cannot differentiate between ground and sky, walking over the edge of a cliff is not an unheard-of occurrence. Suddenly, and without warning, a strong wind blew and catapulted the thick mist clear, revealing that I was nearer the edge than I reckoned, though far enough away for it not to concern me. Lying to the south-east, across a vast opening and expanse of cliffs and cloud, loomed the unforgettable vision of Sgòr an Lochain Uaine and Cairn Toul, two 4,000-foot hills that adjoin the Braeriach massif. Spindrift whipped upwards from their summits like spiralling ghosts, twisting to the music of the gale. Smoke-like wispy cloud was thrown over the edges of cliffs at high speed. A memorable ski tour around the three high hills followed, with no other soul in sight. The ski descent back down to the glen floor, on deep and uniform spring snow, was the best I had experienced in my short skiing career.

Nearby, and lying buried under deep snow that same day, were the plateau's best-known curiosities: the Wells of Dee. These springs are unique in Britain. From an unknowable source, deep within the rock, various small spouts emerge within a few hundred yards of each other, then conjoin and begin their journey to the sea at Aberdeen, many miles distant. The immediate landscape around each individual spring is a sea of colour as the water issues from the ground. The primitive mosses and liverworts that nestle down in the small hollows depend on this unfailing supply for nourishment. Few other plants could survive at such an altitude and in such difficult conditions. It is the highest source of any watercourse in the British Isles, well above 4,000 feet.

After these individual strands of the infant Dee have fused together, the young – but already impressively sized – river then must find an exit from the plateau, which it does, suddenly, by tumbling as a waterfall over cliffs and into Garbh Choire Dhè – the rough corrie of the Dee.[1] This beautiful cataract, known as the Falls of Dee, is invisible in winter, buried under untold amounts of snow. Not until late spring does the water show itself, punching a hole through the sugary, receding April snow.

After it regains its form and exits Garbh Choire Dhè, the nascent Dee flows into the very heart of Braeriach's greatest corrie: An Garbh Choire – Seton Gordon's *Garrachory*. Its meaning in Gaelic is simple: 'the rough corrie'. Many such corries across the Highlands are called Garbh Choire, but only this one gets the article – *An* Garbh Choire. It is *the* rough corrie, not just *any* rough corrie.

The sheer scale of this place is difficult to describe satisfactorily. Even when a walker is in its proximity in clear weather – a relatively rare occurrence – the cliffs guarding each smaller, subsidiary corrie that run off it constrain the full panorama. The summit of Ben Macdui, on the other side of the huge glacial pass of the Lairig Ghru, allows the walker to grasp the full enormity of the Garrachory.[2]

At its south-western extremity lies the highest and most distant subordinate, Garbh Choire Mòr.[3] For every snow-patch devotee, this is our Mecca. It is the place in Scotland – and indeed the whole of Britain – where snow virtually always lies longest.[4] Every trip here in summer and autumn is a mini-adventure. One could almost say it is like a pilgrimage. It is the most isolated place in the Cairngorms, and because of this it is rarely visited, even today. It is fitting, then, that the most difficult place to get to in the Cairngorms is the one that harbours the longest-lasting snows. Only those who harbour real desire will reach it. This is the place where, as a teenager, I saw Muriel Gray on TV standing on a rocky spur looking down upon the snow.

Over the last few years the semi-perennial snow patches at Garbh

---

1   Garbh Choire Dhè is pronounced as 'garra-chor YAY'. The Ordnance Survey incorrectly styles it as *Garbh Choire Dhaidh*.

2   Lairig Ghru is pronounced as 'lahrig GROO'.

3   Pronounced as 'garra-chor MORE', with the emphasis on the third word.

4   Only in 2018 did any snow last longer than here. This was at Aonach Beag, near Ben Nevis.

Choire Mòr have gained a level of exposure they have not previously seen. But it would be wrong to say that they had hitherto been ignored or even unknown. Indeed, from the 1700s until the early 1900s game-keepers and stalkers of the area passed word down, quietly, from parent to child that the last vestige of snow was never known to disappear at Garbh Choire Mòr.

Other pioneering walkers mentioned it as well. In 1829, the writer John Hill Burton (biographer of David Hume, the well-known Scottish philosopher) walked on the summit of Ben Macdui in summer and observed the snow from across the Lairig Ghru. Describing the scene towards Garbh Choire Mòr he saw:

a long wall of precipice, extending several miles along the valley of the Dee. Even in the sunniest weather it is black as midnight, but in a few inequalities on its smooth surface, the snow lies perpetually.

This last word, 'perpetually', occurs sometimes in obscure outdoors literature when used to describe Garbh Choire Mòr's snows. Prior to 1933 the epithet 'perpetually' would have been justified. No man or woman for many generations would have seen this location (and by extension, Scotland) entirely bereft of snow. But this situation wasn't to last forever. The Little Ice Age, that long, cold period in Europe when snow fell harder and for longer, had breathed its last several decades back. So, one warm, sunny September afternoon at Garbh Choire Mòr in 1933, the final snow patch in Scotland could withstand the heat no longer and expired. This hardened piece of post-glacial history, this deformed ice, hundreds of years old, dissolved in the warm air of a Cairn-gorms summer with no one present to give a eulogy or toast its passing.

Nevertheless, it is pleasing, at least to me, to know that despite its quiet, unwitnessed departure, the earnest members of the Scottish Mountaineering Club did not let the snow's disappearance go un-heralded. After they had written a note to various estimable publications, the story gained national interest and was reproduced around Britain. Even the *Belfast Telegraph* of 2 January 1934 ran a piece:

In Scotland there has not been so hot a July since 1903, and a feature of this intensity of the 1933 summer heat was recorded in the 'Meteorological Magazine' which stated that the heat last summer caused all the Highland snowbeds to succumb to its thermal power. This is regarded as something unique, never having occurred in living memory before … [T]he Garbh Choire snowbed under Braeriach – the 'rock where the snowflake reposes' … vanished for the first time in memory, so that by October 1 Scotland was entirely without snow.[5]

So unusual an occurrence was this melted snow that none of the mountaineers or people who frequented the higher Scottish summits expected to see such a thing happen again. And for twenty-six years after 1933 they were to be correct. Normal service resumed, and the snowbeds took on their customary position. That is, up until 1959. By then, more people were in the hills and more eyes were watching. None were sharper than Sandy Tewnion's. By day he earned his living as head biology teacher at Dollar Academy, but during holidays and weekends he was a hill man. On one such day of free time, Tewnion walked across heather and rock to Garbh Choire Mòr to climb one of the corrie's many cliffs. Climbs all over the Cairngorms were being pioneered by an increasing number of people who had more leisure time and an inclination to explore. Tewnion was one such figure. Even at this early date in 1959, 22 July, he worried that the 'perpetual snows' were in trouble once more. Pictures taken by him on this date show his worry was well-founded. Just four patches remained, none of which were large. With another complete disappearance seemingly inevitable, a return visit was necessary just to confirm this. Therefore, on 13 September, Tewnion returned. Expecting the worst, he was not to be surprised. He walked to the lip of the hollow that held the last patch, only to find it empty. Quite probably the first person in hundreds of years to do so, he stood on the bare ground which had been exposed by the second melting this century. No record exists

---

5   *Belfast Telegraph*, 2 January 1934.

of anyone having done so in 1933. Perhaps he was the first person *ever* to do so. It is quite a thought.

As befits his scientific background, Tewnion took various photographs of what he found, in colour. These offer us a unique insight into what was contained in the hollows that had only been exposed to the elements for a matter of weeks in the last several hundred years at least. The spectacle was extraordinary. Where the very last patch had lain, a kind of primordial black residue coated the ground and rocks. To the untrained eye it looked as though an oil tanker had shed some of its load, such was the darkness of the substance. We shall never know what it contained, though it was likely to be many years' worth of decayed grass, mosses and ptarmigan dung that had found its way on to the snow and had become trapped. The melting snow had liberated it on to the boulders that had been covered for so long.

Notable too was the lack of any form of life that surrounded the snow-patch site. The cliff faces here retained their virgin-pink-granite hue. Not even the most antediluvian life form such as lichen or moss could or can establish itself here, so short is the growing season. As a result, the rocks of the cliff face remain blemish-free, as though they've been jet-washed.

Few people noticed all the snow melting this time round. Though it was noted, no letters were despatched to the newspapers. A few people jotted down in their diaries that it had gone, and quietly went about their business otherwise.

The new winter snow that filled the hollow in which Tewnion stood in September 1959 arrived just a few weeks later, in early October. There it lay, undisturbed, for the next thirty-seven years. Normal service had once again been resumed. During this period the two longest-lasting patches at Garbh Choire Mòr acquired names. Not official names as such, but ones bestowed on them by people who were apt to visit these places. It wasn't due to a desire to anthropomorphise the snow, because such things would have been an anathema to the sober members of the Scottish Mountaineering Club. It was a way, rather, of differentiating between things that would otherwise necessitate a convoluted identification system. The most durable patch of all was given the title of

'Sphinx', after the rock climb directly above it. First completed in 1924, this route gained its name years later, in 1942, by W. Thomson Hendry. Apparently, some of the rocks resembled that most ancient of sandstone carvings in Egypt. Today the Sphinx is the best-known snow patch in Britain.

The second longest-lasting, just fifty metres away from the Sphinx, took its name from the gully that starts directly above it. Tall pinnacles of rock soar from the gully's northern wall, so it seemed only fitting to nod to this prominent feature. By extension, the patch took the same name: 'Pinnacles'. Although separated from its near neighbour by no more than a stone's throw, Pinnacles survives much less often. The fact that two patches of snow located in virtually the same place can have such different outcomes is an oddity.

The Sphinx and Pinnacles patches' longevity is very much predicated on the direction of the prevailing winter winds. Moisture-laden Atlantic storms can blow massive quantities of snow into Garbh Choire Mòr, and Pinnacles Gully acts as a conduit, funnelling the snow down through cliffs until it reaches the hollows that the two patches occupy. We can only speculate as to the depth of the snow that can accumulate here in exceptional years. 'Mind-boggling' would cover it, though sixty to seventy feet if one prefers more precise measurements. In such snowy winters the survival of the Sphinx and Pinnacles patches through the entire year is almost assured, even from a very early date in spring.

From 1959, when all the snow melted for the second time, the run of winters thereafter was generally snowy. 1963, 1967, 1975, 1977, 1979, 1983, 1986 and 1994 produced lots of it. Others such as 1972, 1987, 1990, 1991 and 1993 had only slightly less.

However, in 1996 a seemingly seminal event occurred. The preceding winter of 1995–96 was not vintage for snowfall at all. Much of what tumbled from the sky did so on southeasterly winds. This type of weather system may benefit some ski areas, but it is much less useful at filling up the east- and north-east-facing hollows that hold the most durable snow. As a result of the lack of southwesterly storms, the relatively shallow snow patches melted, and quickly. On 1 October 1996 the Sphinx and

Pinnacles hollows were visited, and the latter had gone, with the former critically small. It vanished utterly a week or so later. The reason this event appears to be seminal is because of what came next.

It would have been reasonable in mid-October 1996 to conclude that, given it had occurred only three times since the 1700s, another complete vanishing of all snow in Scotland would not happen for many years to come. After all, the most recent disappearance before 1996 was thirty-seven years previously, and before that twenty-six years. But this was to be quite wrong. Thirty-seven years had elapsed between 1959 and 1996, but we had to wait a mere seven years for the next one. Not only that, 2003 saw the Sphinx expire by the almost unimaginably early date of 23 August. To be fair, the summer of that year was as good as any I can remember. Months of warm weather and bright sunshine dealt a death blow to the meagre depths that had accumulated in a poor and mild winter. Just seven years now separated the last two vanishings. Could it get worse?

The next two years after 2003 were touch and go, too. In 2004 only the Sphinx survived the year in Scotland, and in 2005 just it and Pinnacles. The run of extremely poor years continued into 2006 when, yet again, all snow vanished. This run was unprecedented.

Alarm bells by now were starting to ring out. The once 'perpetual' snows of Garbh Choire Mòr had succumbed to the weather three times in just eleven years (1996, 2003 and 2006), having done so only twice in the previous 300 (1933 and 1959). The media were running more and more stories on what at the time was called 'global warming', although we now refer to this as 'climate change'. Climatologists suggested that worldwide temperatures were on the up, and that glaciers were in retreat. The same phenomenon was being put forward for Scotland's vanishing snows. Ski centres were in financial difficulty, and there seemed to be a correlation between each poor trading year and how much snow survived to winter. Both were in retreat.

But then, a reprieve. Garbh Choire Mòr's Sphinx and Pinnacles both rallied in 2007 and were big when visited in late October of that year. Several other patches of snow survived in Scotland for the first time in a few years. The story was repeated in 2008, with even larger remnants in place by the time they were buried by winter's new falls. Indeed, good quantities of snow persisted at Garbh Choire Mòr for the next few years, culminating in the outrageous volumes which could be seen in summer 2015. In that year, it was more to do with the fact that spring seems to have been entirely bypassed, rather than any exceptional snowfall during winter. In April and May there were no appreciable thaws, and winter refused to go. During the first week in June the high Cairngorms looked as though they were still in deepest midwinter, and Garbh Choire Mòr was a massive, snow-filled cauldron.

I recall walking there on 13 September 2015, scarcely able to take in what I was seeing. The cliffs of the corrie headwall had at their foot a massive wreath of snow that stretched unbroken for hundreds of metres. By this stage of the year the individual patches along the cliffs would normally have formed, but on that day they were still conjoined, and showed no inclination in splitting any time soon. The quantity of snow that survived that year was more than any other since 1994 and deserved to stand among the very best years of the last hundred. Proof that you write off the Scottish winter only at your peril.

Or do you?

The dark days of the mid-1990s and 2000s had passed, and a new spirit of optimism had emerged on the back of a series of decent winters. Maybe this period had been an aberration. A mere blip. Unfortunately, it wasn't, because the years 2017 and 2018 would be unprecedented.

All told, I went to Garbh Choire Mòr five times in 2017, and saw it numerous times from afar. Each time I visited I could see the Sphinx and Pinnacles reduce. I hoped that heavy snow would come to their rescue and bury them, ensuring their persistence for at least another year. But, alas, no snow came. On my last trip there that year, on 30 September, the snow was so small and light I could lift it up. It felt odd to be holding this lunch-platter-sized piece of old, dirty snow that had fallen from the sky late in 2006, eleven years previously, the oldest relic of its type in Britain. For some divine amusement, the weather gods laughed at us as we stood by the dying patch, sending a few derisory flakes downward. For a minute it got quite heavy, but then it stopped just as it looked as though it might do something meaningful. So it was, then. Another year, another complete disappearance. The tally now stood at four since 1996.

The next year, 2018, brought a dose of *déjà vu* with it. Worryingly little snow fell in winter across the Cairngorms, and at the start of August the Sphinx and Pinnacles were looking sickly and feeble, as though diseased with some form of illness. Numerous visits were undertaken, and I managed to rouse enough interest in others so that the burden of getting there was shared. But it made no difference how many people went there, as the result remained the same. Each set of photographs showed the Sphinx and Pinnacles reducing in size. By mid-September I had written them off. Curiosity got the better of me a couple of weeks later, and almost a year to the day since I went in 2017, I ventured back on 29 September 2018.

Clambering up the stony ground I knew it wouldn't – couldn't – be there, but always there is a modicum of optimistic hope that, somehow, it has clung on. On that September day, as I peered over the hollow where the snow usually lies, I saw nothing. Not a lick of anything white. Here I was, almost sixty years after Sandy Tewnion, standing in exactly the same spot and seeing exactly the same thing – black matter deposited on bare rock. It then occurred to me that I had been one of only a very select few who'd stood on this hallowed piece of ground. It felt like a little comfort, but my overriding feeling was sadness. For the first time in recorded history the Sphinx had disappeared two years in succession. When I brought news of this disappearance to the attention of the wider scientific community, it attracted – just as in 1933 – the attention of news agencies not just in the UK but around the world. 'Scotland's glacier melts' was a common headline. Preposterous, of course, but the level of interest in the story had been quite remarkable. I spoke to radio stations in Canada and New Zealand, where both countries have enormous reserves of glacial ice. It seemed incredible to me. Why the interest? Probably it had to do with the heightened awareness of climate-change issues, where parochial bellwether incidents like these attracted attention out of all proportion to their scale. For a time in 2018, the Sphinx was the most famous patch of snow in the world.

* * *

Garbh Choire Mòr is a unique environment. In large part this is to do with its remote location. It is the most isolated corrie in the Cairngorms, and just getting to it, never mind the snows, is an undertaking in itself. Any ambitions to see it need to be worked around a calendar, with the best time of the year to visit being early September. The days have sufficient light in them to allow a fairly steady walk there and back in a day from wherever one wants to start. The weather in September can often be agreeable, but as ever with the Cairngorms a large dose of salt needs to be taken with any such recommendation.

I have, to date, been to the corrie in excess of twenty-five times, and by every conceivable route. I have tested all approaches, each slope, many gullies, looking always for the easiest way in and out. I settled long ago on there being no such thing as 'easy' when talking about getting to and from Garbh Choire Mòr. From every angle it is a long, hard and de-manding place to attain. If venturing there from the plateau above, with the intention of dropping in, there is only one chink in its armour. It is an unnamed gully with enough of a slope as to permit the walker down-wards access. If this gully were a degree or two steeper it would be impossible to descend. As it is, loose stone and moss abound, with each footstep a lottery. A forward fall would be catastrophic, ending almost certainly in a serious injury or, God forbid, worse. When in pairs or threes, one walker must go ahead and then stand well to the side while the others come down. The whole gully is a ticking time bomb of loose rock, ready to be dislodged by one misplaced boot. Even a medium-sized boulder set off just ten yards up could knock a lower walker flying. And this route is my preferred one! As with most gullies and steep slopes, though, exiting is a far easier and more pleasant experience than dropping in.

The safest way of getting into the heart of the corrie is via the Lairig Ghru. Only from that magnificent glacial trough is there an absence of any steep ground. Once off the indistinct track that runs through the Lairig Ghru, the walker will turn to their right if approaching from the north, into the huge opening of An Garbh Choire – the Garrachory.

At this point, Garbh Choire Mòr is visible in the distance, at the terminus. It appears laughably distant.

The walker is then obliged to negotiate ankle-breaking peat hags, cunningly disguised by thigh-height heather. Stout footwear is essential here to stop one's feet getting wet from the interminable bog that besets every person who ventures here. After two miles of this terrain are disposed of, there comes the not insignificant matter of fording the embryonic River Dee. In spate, crossing it is next to impossible. Sticking to the north bank is the safest option, but by then most walkers are so beguiled by the terrain, and so tempted by the sight of the An Garbh Choire refuge on the opposite bank, that crossing the stream becomes an imperative.

When standing at the refuge, with perhaps a flask of tea consumed and breath regained, one is right in the heart of the corrie. It is a sight of sights. Coire Bhrochain and Garbh Choire Dhè are prominent, with ancient granite cliffs seemingly encircling the walker. Garbh Choire Mòr seems no longer to be unobtainable, being only about an hour or so farther on.

When one enters Garbh Choire Mòr – the Garrachory's innermost sanctum, and the home of the snows, something *feels* different. Coming in via this route has an almost religious quality to it. Because all the while one has been walking towards the snow along a walled corridor of granite on each side, it has the feeling of walking through a massive cathedral, towards the altar. In this instance the altar is the uppermost cliff of Garbh Choire Mòr. The innermost holy shrine is where the snow is contained. The pilgrims come to it, and the process of getting there feels like a purification. The final push up to the foot of this cliff is the final penance before absolution.

Arriving at the snow has been described by some as anticlimactic. Because such effort has been expended in reaching the Sphinx or Pinnacles, the sight of a twenty-foot-long piece of dirty snow can apparently be underwhelming. I must confess that I have never felt that way. I savour each trip to this location, as there is always so much to see and hear, to experience. The snow is never quite the same, and a visit is often accompanied by the cruck-crucking of a raven above, or the comb-across-paper rasp of the ptarmigan. The mellifluous warbling of the snow bunting is frequently heard in the misty gloom of the summit cliffs, invisible to the flat-footed, transient visitor to their domain. Carcasses of unknown animals are dotted around, often killed either by falling from above or being caught in an avalanche. To some people this place can feel almost malevolent: a giant cul-de-sac of intimidating proportions, a cold dead end.

No visit to An Garbh Choire is underwhelming. It is a place that sears itself on the memory. Because of its isolation it will never be a place of popular attention but, because of this, it will remain forever the most special of places. No single location in Scotland that I have been to carries so much beauty within it, nor as much magic or shades of intrigue. It may no longer be the 'place of the perpetual snows', but it is a place which no one who has been to can forget as long as their memory holds:

> ... the Ridge of Braeriach, said, by an eminent calculator of altitudes, to have 2,000 feet of sheer precipice; that 2,000 feet of precipice is the object which it now almost aches your eyes to look upon – a flat black mass, streaked with snow, and sometimes intruded on by a cloud which divides the upper regions from the lower. It is probable that now, in mid-day, a hot sun gilds its black front, and mocks its streaks of snow, while a dead, unearthly silence pervades the mass.[6]

---

6   *North British Daily Mail*, 2 October 1848.

five

The back of beyond

Over the years, I have made innumerable journeys into the hills to chronicle patches of snow. I am fortunate to have seen a good deal of the country and ventured to many places that others would have little reason to go to. Many of these trips have been uneventful, save for the snow itself. Quite a few, though, have been a combination of nerve-wracking, funny, terrifying, and sometimes all these things. Some of the most eventful and memorable are described below.

* * *

29 August 2017, Garbh Choire Mòr

'Are you sure he understands what's involved in this?' I asked.

'Oh yes, he's a keen hillwalker and is used to big days out,' replied the editor of *The Great Outdoors* magazine.

'You told him about the rough terrain, and the long bike haul just to get to the start of the walk?' I added.

'Definitely. We're hiring a bike for him tomorrow and he'll meet you at Whitewell car park at 9.30 a.m. as agreed. He's looking forward to it.'

I hung up, satisfied that my companion the next day was sufficiently well briefed on the trip he wanted to go on.

It was about a week previously that I'd received an email from the editor of *The Great Outdoors* magazine asking if the well-known Irish comedian Ed Byrne could accompany me on a trip to see the by now famous snow patches of Garbh Choire Mòr. He found it fascinating that people had such a passion, rather than, say, collecting football shirts. He wrote a monthly column for the magazine and was always on the lookout for unusual outings to file copy on. He had most certainly come to the right place.

We met the next morning as planned at Whitewell car park, near Aviemore. The weather was overcast but dry. (Any day that starts dry is a blessing. It can be demoralising to start a long journey in full water-proofs.) After some pleasantries we set off on the first five and a half miles of the day, which were to be done on two wheels. We started up the well-constructed track towards Glen Einich. All was well as the first couple of miles went by, but then the road started to steepen. It was at this point that I had my first suspicions that we could be in for quite a testing day. We were obliged to dismount from the bikes and push them for several sections of the journey in. Conversation was necessarily brief and only then when information needed to be relayed, such was the effort expended. But, at last, after about an hour of huffing and puffing into the wind, we arrived at the spot where the bikes could be set aside for the journey back. I assured Ed that the return downhill was well worth the effort of the tiring inward cycle.

If he thought the next section would offer any respite, this was quickly dispelled by the sight of what lay ahead. To attain the summit plateau of Braeriach we would need to follow the long line of the Allt Buidheneach, a small stream that gallops down from about 3,500 feet. In winter and spring, the gully that this stream empties is known simply as 'the escalator' to ski tourers, as it offers reliable and straightforward access upwards at a constant (steep) angle. At times of the year when snow is not present, like that day, it's an unremittingly testing upwards trudge through high heather and slippery grass. It is, though, the quickest and most direct route up. I thought that Ed, being an experienced hillwalker, would make short work of it. Perhaps I was just used to the terrain, or had lots of experience of this particular gully, but in the end the upwards traverse proved to be rather trickier for Ed than I had envisaged. In fact, during the walk up he said, (half) jokingly, 'You know, Iain, I'm starting to go off you!'

At the top of the gully we paused for a bite to eat and caught our breath. It had been a lung-bursting ascent after all. Once refuelled and refreshed we made our way across the plateau. With easy ground in front of us we enjoyed decent progress and had time to engage in some chit-chat. The cool late-August day kept the temperature pleasant. Things were starting to look up. That is, until two miles later when we reached the drop-in point to Garbh Choire Mòr.

As described in the previous chapter, the gully that affords the quickest and safest ('safe' is a relative term when talking about Cairngorms gullies) access is a steep defile where what little solid ground that does exist is never more than a misjudged footstep away from disintegrating into a welter of exploding gravel. We negotiated the gully with great care, taking over twice as long as I usually would. Such care and time are absolutely necessary when more than one person is going up or down a gully. On the way down, the Sphinx and Pinnacles snow patches came into view. Despite my assurances that they weren't as far away as they looked, Ed affirmed to me afterwards that they were.

With the gully safely negotiated, all that was left to do was to walk to the snow and do the necessary: measure and photograph it for posterity. When we arrived at the two white remnants I could sense that Ed found the whole thing bordering on the absurd. Here we were, about four hours away from relative civilisation, down a stony chute at the foot of crumbling cliffs. Beside us were two melting patches of snow that were barely thirty feet long. Though there was a grudging admiration from him for the dedication that people like me displayed, it was just too esoteric for him to appreciate fully.

All that said, he was genuinely interested in the Sphinx and Pinnacles patches. That such things could exist in late August, at the tail end of a long summer, seemed to intrigue him. I got him to pose beside the Sphinx so that we could take some photographs for his piece.

'What way are we going out?' Ed asked me when we were set to leave.

'Back the way we came in,' I replied.

'Isn't there another way out?'

There are several ways in and out of Garbh Choire Mòr, but all of them involve varying degrees of risk. When Ed asked me the question, all these routes flashed through my head. I discounted Great Gully immediately. Though it was in close proximity to us, two people going up there would be asking for trouble. It is a ticking time bomb of loose rock which is held together – if at all – by moss and gravel. We would need to be glued together going upwards so as to avoid one of us sending a rock hurtling down and taking the other with it. Great Gully is too narrow for close-quarter climbing in places, so that was out. The other major gully in the corrie, between Sgòr an Lochain Uaine and Garbh Choire Mòr, I discounted for the same reasons. I'd been in and out of them both before, but I vowed the last time never to do so again. The third option, leaving via the Lairig Ghru, would necessitate such a diversion that it would mean us getting back in the dark, even if it were easily the safest way out across uncomplicated ground.

'The only real option,' I said at last, 'is Pinnacles Gully.' I pointed directly upwards to the improbable gouge in the cliffs above. 'I don't really recommend it, though.'

'Have you done it before?' Ed asked.

'Several times. It's doable, but as I said I don't recommend it as it's a bit scrambly.'

'Let's go for it,' said Ed, and so off we went. He seemed confident enough to attempt this exit, so I assumed he was no stranger to a bit of sport.

The first section of the gully was relatively straightforward. The terrain, steep but not overly so, offered only a little difficulty. There were good handholds and no real tricky sections. Before long, however, the gully kinked to the left and the going became sportier. We encountered a couple of cruxes, which with some jiggery-pokery we managed to negotiate. All the while, though, I had to be vigilant with loose rock. With me taking the lead, I needed to ensure that each upwards movement was on as good a standing as possible, lest I send one of Britain's best-known comedians hurtling to a sticky end hundreds of feet below.

The 'good standing' I searched for at each turn was not always easy to find. The ground in this soaking defile is a constantly changing melange of splintered rock, moss, grit, football-sized boulders and streaming water. Booby traps lie in constant wait, requiring only one misplaced hand or foot to rend a boulder asunder.

But, by good fortune and basic hill competence, we were nearing the top. The end was definitely in sight. However, there was one final chock-stone to negotiate, and it was – to use the words of an acquaintance of mine who coined it when we did the same gully – 'a beezer'. It blocked the entire gully and had a slight overhang. Remembering it from the previous times I had ascended, I recalled what move to employ to circumvent it. Once I was up, Ed tried to do the same. Over and over he contrived to pull himself up, but each time the rock defeated him. He prodded and probed with both hands and feet, trying desperately to get some traction, but nothing worked.

'How on earth did you do that?' he asked, looking up at me from his precarious position.

'I'm not sure. I just sort of manufactured a handhold and used my legs to crawl up,' was my unhelpful response. Once again he tried, and once again he was unsuccessful.

There was no possible way of going back down, for to do so would be akin to signing our own death warrants. Going up a steep gully is one thing, but descending it is another entirely. Upwards momentum is relatively easy to control because gravity constantly checks it. Downwards momentum has no such check. If one were to lose any sort of footing or handhold, and in this place there were many only too happy to give way, the outcome was likely to be grave. It would probably involve moving downwards at great speed until halted by a large rock or some other stony feature. I need give no explanation as to the associated medical repercussions of such an event.

It was at that point I thought what an embarrassment it would be to have to call Mountain Rescue. Braemar Mountain Rescue follow me and Ed on Twitter. Having to explain to them why we came up this preposterous route would be the outdoors' equivalent of a trip to the headteacher's office. It was unthinkable. So, there was nothing else for it. I had to do the honourable thing.

'Grab my hand. I'll haul you up,' I said, thrusting my arm downwards. In order to secure myself I wedged my right foot into a good crack in the granite, and my left against the chockstone blocking Ed's escape. My knees and one of my elbows had a decent placement on the rock, allowing for good purchase. It must be said at this juncture that the method of extrication I am about to describe is absolutely discouraged. Hauling someone up by their hand, when a slip would mean the direst of out-comes, will be found in no guidebooks and nor will it be taught in mountain leader courses. It was born of absolute necessity.

'Are you sure?' replied Ed, looking slightly worried. I assured him that my stance was sound, and that it was either this or calling 999. Resigned, and perhaps anticipating unfavourable headlines, he grabbed my hand using the 'monkey grip' (that lock where fingers are curled round the fingers of the other hand, the safest method of ensuring no slippage). With an almighty pull I heaved him up until he was able to wedge a foot into a wide crack in the rock. This, in turn, allowed him to obtain suffi-cient purchase on another rock so as to permit upwards momentum. At last! We had negotiated the chockstone and we exited the gully. In the interests of decency, I shall omit the industrial language that both of us used when we emerged on to the plateau once again. It was a huge relief to finally get back on to some flat ground. Alas, for Ed the trial wasn't yet over. The downwards 'walk' was done in knee-high heather which often concealed boulders that were apt to twist the ankle. Many, many curses later we were back at the track and bikes.

The return to the cars on two wheels was, as it always is, a joy. A tail-wind pushed us back down the hill, with very limited effort required. It took barely twenty-five minutes, and we parted after loading the bikes on to our respective cars.

I read Ed's subsequent write-up in *The Great Outdoors* magazine a month or so later. It was clear from his piece that he is unlikely ever to join the ranks of us who venture into the high hills to look for snow, but there was a generous testimonial. He also said that this was one of the hardest days he'd ever had on the hill. Secretly I was rather pleased at this admission.

* * *

25 October 2015, Beinn Bhrotain

I had long lost count of the trips I'd made into the hills that year. Twenty? Thirty? Who knows?

Spring 2015's obdurate refusal to pass on the seasonal baton had left the high peaks and passes of the Cairngorms looking unusually snowy, and it was now early summer. The frightful cliffs that rise vertiginously from many of Braeriach's corries had upon them a full winter jacket, yet it was now June. At the same time, the pregnant-like summit bump of Ben Macdui rose white and unblemished above everything else within sight.

Summer passed, though. Much of the snow melted, of course, but relics of winter and spring remained across Am Monadh Ruadh – the Gaelic name of the Cairngorms – into September and even October. It was no exaggeration to say that the scene was a facsimile of one from many autumns past. John Taylor, King James VI's 'Water Poet', who journeyed to Braemar in 1610, would have recognised it:

> There I saw Mount Ben Avon, with a furred mist upon his snowy head instead of a nightcap: (for you must understand, that the oldest man alive never saw but the snow was on the top of [many] of those hills, both in summer, as well as in winter.)[1]

In a normal year   so far as any year in these parishes can be thus described – counting the number of patches of snow that endure to the tenth month would be an exercise able to be conducted over the course of a weekend. This, however, was not such a year.

For some time leading up to the start of October, my Saturdays and

---

1   Taylor, John (1618), *The pennyles pilgrimage, or the money-lesse perambulation, of John Taylor, alias the Kings Majesties water-poet*, London: Printed by Edw. Allde, at the charges of the author.

Sundays, as well as a good chunk of my annual leave entitlement from work, were given over to what sometimes felt like a never-ending circle of repeat visits to the inaccessible nooks and crannies of Scotland's highest tops, gathering data for the annual snow-patch paper.

24 October saw one such visit. That day's target was the immense bulk of Beinn Bhrotain, 'the hill of the mastiff'. Like so many of its Cairngorms brethren it has a whale-like appearance from distance. Traverse around its northern flank, however, and the rolling countenance is brutally sliced open by the shattered and splintered granite cliffs above Glen Geusachan. For me, no lover of heights or steep cliffs, my path was to be more benign.

Cycling from the spate-engorged Linn o' Dee at daybreak, I made for White Bridge. Though progress on two wheels would have been easier on the east side of the adolescent River Dee, its fording would have been impossible, given the quantity of rain that was now trying solemnly to get back to the sea. Wiser counsel suggested the western approach, crossing White Bridge over the Geldie – which swells the Dee to double its size – and on towards the foot of the mastiff hill.

Arriving at White Bridge I paused briefly and marked the clouds that were lifting. The mature orange and browns of the now dormant autumn vegetation were in stark contrast to the gleaming white of Ben Macdui and Braeriach, whose top 500 feet were resplendent in a castor-sugar covering of fresh snow. It was then, also, that I caught my first glimpse of Bhrotain's snowy white spot. It was exactly as reported: sitting in the upper reaches of Coire an t-Sneachda, 'the corrie of the snow'. The Gaels were, apparently, noting long-lasting snow locations hundreds of years ago.

Onwards.

But, alas, not for long. The normally placid Allt Iarnaidh, which drains but a small area of the southern slopes of Beinn Bhrotain, was a seething, foaming torrent of angry water. Luckily, just upstream, its course was constricted by a narrowing of the gully, and a simple hop over with the bike was sufficient to overcome what would have otherwise been an impassable barrier.

Ten minutes or so later, I was at the hardly discernible start of the path which led up the course of the Allt Garbh. This handsome brook reached upwards right into the heart of the hill, emanating directly from the snowy corrie that I was aiming at. For the next three miles or so it would be my noisy but unwavering companion.

The terrain was in no hurry to lend me height. A slow and steady upwards march through thick, tussocky grass and heather necessitated close alignment to the chatty burn that cascaded over the virgin granite outcrops, stripping anything unlucky enough to grow within its cold reach.

Eventually, some two hours after parking the bike, I reached the corrie. Last year's patch of snow sat in a large hollow just below the horizon's edge. Now, on easier terrain, I made for it, noticing a large inverted 'V' carved in its southern edge. I knew immediately what this meant: a tunnel. But this was surely no ordinary tunnel. It was one that had been months in the making. Water and wind had carved it out during the short summer and autumn. I hurried towards it across good ground. Two ptarmigan took to the air, their winter coats of brilliant white not yet developed.

Upon reaching the opening, I peered in excitedly. Snow tunnels are one of the great sights of the Highlands, and I had been at enough of them to know how wondrous they could be. Even so, seldom had I seen anything like this one in Scotland. I crouched motionless, then dumb-founded. A cold wind, far cooler than the ambient air temperature, passed down through the tunnel and across my face. A coldness that had been laid down some eleven months previously was blowing away with the wind.

Mighty pillars of white snow on each side of the small burn supported the most stunning of arches, which ran the full length of the seventy-five-foot-long patch. Translucent blues and whites, caused by thinning snow being pierced by the daylight, lent an otherworldly quality to what would normally be just one of a million anonymous watercourses. The tiny stream's babbling noise offered a lyrical soundtrack to the atmosphere.

In addition to the colours of the snow, mosses of luminous green hugged the waterside. Large, pink granite blocks also contributed to the kaleidoscope of colours, which were almost overwhelming. Today, the small stream was not anonymous: it was the most beautiful place in the whole of the Cairngorms, and probably the entire country. At least, to me it was.

I thanked whichever of the responsible gods that, for this brief window of perhaps a few days, I was fortunate enough to be there and see the snow before it collapsed on to the stream. Its blocks would dissolve and end up feeding the River Dee, as all precipitation on this hill must ultimately do. In my head I drew parallels between this scene and when blossom appears on the trees in Japan: an ephemeral window of staggering beauty which disappears as quickly as it comes into being. I drank it all in, unsure if I would ever come to this location again and witness such an exhibition of colour and beauty.

* * *

19–20 July 2019, Ben Nevis

All told, there were eight of us. Initially there had been seven, but a last-minute call went out on social media to recruit one more Sherpa. The amount of kit we'd need over the course of the next day and a half was significant. Luckily, the internet did its thing and a local guide, suitably remunerated, was roped in for the expedition. We all met up at Ben Nevis's north face car park, near Fort William, and introduced ourselves. It was good to finally match faces and names.

About two months previously I had been contacted by a film crew in Wales with a view to doing a piece for TV on the north face of 'the Ben'. The producer had seen some photographs online of the snow tunnels on Ben Nevis and thought that a visit there would make for superb and unusual visuals. She aimed to wrap these visuals around some commentary on my snow-patch work, and how that fed into the wider climate-change debate. The piece was to be broadcast on national TV.

The plan for the trip, previously hatched, was to make for the Charles Inglis Clark (CIC) memorial hut that is situated at the foot of Ben Nevis's towering cliffs. It is the only alpine hut in the UK and serves as a superb base in winter for climbers not wishing to make the six-mile round trip on top of an already long and exhausting day.

As we chatted in the car park on that Friday afternoon the weather was mild and cloudy, with a hint of moisture in the air. There was also very little wind. A mere breeze, in fact. Readers familiar with the climate of Scotland's west coast will be unsurprised to learn what came next: midges. And they were on a scale that I had seldom seen, even in the insect paradise that is Lochaber. Thick black clouds of them hung in the air so profusely that they appeared to darken the very light. No one was to be spared. They attacked us indiscriminately and remorselessly. We decided almost immediately that any chit-chat ought to be confined to the walk up to the hut, lest we be bitten half to death by these tiny, though indomitable, bloodsuckers.

Our group consisted of me, the presenter, a producer, a cameraman, a sound man, a Sherpa and two guides. On the walk up I started a conversation with the presenter, Lucy. She did weather forecasting on local and sometimes national TV. A serious though not overly restrictive disability meant that she had the use of only one hand. I did think prior to meeting up that this might be an issue, given where we were going and the nature of the terrain. But my fears were to be absolutely unfounded.

The path that led from the car park to the CIC hut was relatively recently constructed and in excellent repair. Good progress was made and by teatime we were safely ensconced within the solidly built granite walls of the hut. To prove to the group that I was more than just a one-dimensional snow enthusiast, I served all eight of us up a homemade vegetarian chilli that I had made the previous evening. (I surmised that because everyone woke up the following morning without complaining of poor health then it was a reasonable meal.)

There was a party of three people already at the CIC hut that Friday evening. A middle-aged couple were accompanied by the woman's mother, who was in her mid-eighties. To our astonishment they informed us that all three of them intended to climb Tower Ridge the next day. This ridge is the longest sustained climb in Britain and is a significant undertaking even for a fit climber. We wished them well and at about 11 p.m. we all hit the sack for as good a night's sleep as eleven people in a dormitory can have. Other than walking boots and waterproofs, earplugs are the single most important piece of kit that I pack in an overnight bag.

By the time we had awoken the next morning, the three climbers had already departed. The weather was not looking too promising, with low cloud and rain threatening. As we all left the hut after a hearty breakfast, we bumped into a local climber who had already walked up from the car park with the intention of also doing Tower Ridge with a group he was leading. We told him about the elderly lady who was with her daughter and son-in-law, and said that they might catch them up on the climb. We said our goodbyes and started on the day's walk.

The idea for this trip originated with the TV producer. She'd seen photographs that were taken by *Guardian* photographer Murdo Mac-Leod after he and I visited the same place back in 2016. *The Guardian* put the best one of that day in the centrefold of their national edition. The photo showed me standing underneath a huge cavern of snow that had been sculpted out by the wind. The shape of this cavern resembled the dome of a cathedral, and it made for a striking image. If, somehow, we could replicate that today, then the producer would be over the moon. I had warned her, though, that because of winter's poor snowfall the chances of getting anything so impressive were small.

Our destination was Ben Nevis's biggest, longest and most notorious gully: Observatory Gully. Though its entrance is only thirty minutes or so's walk from the CIC hut, it is not a place for the faint of heart. Like so many Scottish gullies it is a welter of loose rock and rough terrain. Ankle-breaking boulders and holes are everywhere, and one must be vigilant when walking into the upper reaches. It is also very long. Its jaws open at an altitude of about 2,800 feet and it runs all the way to the summit, 1,600 feet above. To cap it all, the gully is in a constant state of decay. Some of Scotland's highest cliffs lie on each side of the gully and are prone to shedding their skin from time to time. A few times I have been in there and heard invisible rocks scuttling downwards through the mist. There is little doubt that if one of these rocks were to collide with a person then it would be curtains for them. With eight of us up there the chances of being hit were increased dramatically.

Of all the places I visit regularly, Observatory Gully is the one where I have never felt entirely comfortable. The reasons listed above are part of it, but it goes deeper. Whether my impression of it is shaped by events that have occurred there in the last few years I couldn't say for sure, but I think there is at least some truth to that. The gully is full of ghosts, created by tragic events. Many people have in the past met their ends there, often in grizzly ways. Though admittedly the vast majority happen in winter, and to winter climbers, the place resonates with sadness. Even at the gully's entrance, before setting foot in it, one is reminded of one recent incident.

In February 2016, a young couple entered the ravine at its narrow opening point. (Though the gully is massive and wide for much of its length, the exact point of entry is squeezed like an hourglass's middle.) They were likely going up part of the gully to attain the start point of Tower Ridge. Before they got there, though, they were hit by a massive avalanche and swept hundreds of feet out of the gully. Their bodies were found some five weeks later.

The gully is also full of the hill's detritus. In summer, when the snow melts, one can often find climbing gear such as ropes, ice axes, ice screws, gloves, hats, you name it. Each time I pick one of these up, particularly an axe or screw, I wonder if its owner had to be carried off the hill. They are solemn reminders of what this place is capable of visiting on un-suspecting or unfortunate climbers.

Anyway, despite having two guides with us (one of whom, Heather, was Scotland's best known and respected), I was still a tad apprehensive about how individuals would cope once we got into the bowels of Observatory Gully.

At first it was all fine. We kept the pace steady, with me leading through the boulders. I knew well the faint path that wove upwards, having used it dozens of times over the years. On the way up I kept an eye out for our three roommates from the CIC hut but, despite looking at their most likely route, I saw nothing. The rest of our group were in fine fettle, all things considered. Unfortunately, the weather gods were not smiling on us, with cloud coming down and reducing the visibility. This had the added effect of making the gully feel quite claustrophobic. It was at this point I first sensed some uneasiness within the less experienced members of the group.

I watched with admiration as Lucy clambered up the gully. A native of Tottenham in London, she'd never done anything like this before (I didn't know this previously) but, although I could see that she was nervous, she displayed a mountaineer's attitude. Always smiling, she did exactly as instructed by the guides: no mean feat considering she only had the use of her left hand. The sound man and cameraman both plodded up as well, displaying an admirable stoicism in the face of the testing conditions. Of most concern to me was the producer. She was starting to struggle with the terrain as we ventured higher up the gully. The pace had slowed dramatically, and I began to wonder if she was going to make it up to the snows, which were still some 1,000 feet above us. However, she was in good hands with Heather, and I thought that a slow and steady approach would see us get there in the end.

Up we went again after a short break. The ground upon which we walked was unrelenting in its steepness and difficulty. Though in theory the amount of ascent we were doing wasn't too bad, it was inordinately tiring. For each three steps one takes up the gully, one needs to add another for all the backwards sliding that occurs on the loose gravel. Then, about ten minutes after the previous break, we heard a loud cracking noise from above. I knew exactly what this noise meant and so did Heather. It was a rockfall. Admittedly it sounded minor, but with eight people present no chances could be taken. Heather shouted to everyone to get close to a small cliff face that offered excellent protection from anything falling from above. In the gloom we couldn't see where this rock came from, nor where it was going. The noise of it crashing down the gully slowly dissipated. We looked at each other in total silence until we were happy it was safe to continue. The sense of nervousness amongst the group was now palpable.

It became obvious at that point the producer was at the end of her journey upwards. The rockfall was the last straw. Heather agreed to take her back down, as it had – understandably – all become a bit too difficult.

She remained in good spirits for all that and said, smiling through some tears, 'I can't believe you do this for a hobby, Iain.'

Shortly after the producer departed in Heather's good care, we reached our destination. As feared, the lower of the two patches that I thought might give our stunning visuals was in a sorry state. Mild weather over the last two weeks had decimated it to the point where it would be dangerous to enter one of the tunnels. Lumps of snow lay around the entrance to the main passageway, indicating that it was unstable and prone to collapse. Plan B was enacted. This involved a short climb to the longest-lasting patch of snow on Ben Nevis, at the head of the gully. It was in much better shape than the last one, and we got some footage. Not quite the stunning shots of previous years, but enough to justify the extra effort in getting there.

The crew were thankful to now be on the way down. For every step we took, I could almost feel the relief grow amongst them. This was magnified when we descended below the mist, and the views opened up. Thick mist has a peculiar way of distorting scale and distance. If a prolonged period of time is spent in it, then it can demoralise the unprepared walker. This is especially true if the walker is unfamiliar with the terrain. By the time we exited the gully proper, the weather was much improved, as were the spirits of the party.

We arrived back at the CIC hut to pack our belongings and make our way off the hill. Fortuitously we again bumped into the local climber who had been up Tower Ridge. He told us that his group had not encountered the three climbers who had been in the hut the previous evening. I thought this most odd, with only three explanations. Firstly, the eighty-something-year-old climber and her daughter and son-in-law had gone at such a pace that the climber hadn't caught them. The second was that they had looked at the weather and decided to do something else. The third reason was that they had become lost. As it transpired, it was the third reason. We heard the following day that they missed the turn-off for Tower Ridge and ended up going up some other route, got lost, and had to be airlifted off the hill. Our adventure was much tamer by comparison.

On the way down, I reflected on the previous twenty-four hours and vowed never again to take anyone up on to that part of Ben Nevis unless they were experienced hillwalkers. Though the crew had been admirably stoic in the face of some difficult conditions, the north-east face of the Ben is no place for the inexperienced.

* * *

25 November 2018, Aonach Beag

The following section is in note form, taken in part from a diary entry of a visit I made. It describes the thought process of the day: a series of situations and decisions taken in real time.

Glen Nevis. Stop car quickly. 9.30 a.m. Four shafts of sunlight splitting the clouds on the shoulder of Ben Nevis. Beautiful. Take photograph. Stand for five minutes. Three stags spring upwards on the crags to the east. Cloud regroups. Shafts of sunlight shrink and disappear. Back to gloom. Drive to Steall car park at head of glen. Get out of car. Cold. Pack extra fleece. Winter or summer boots? Hmm. Snow sitting on Mamores above 3,000 feet. Winter boots. They're heavy and cumbersome. Don't really want to wear them but OK then. Tell myself I'll regret it if I don't wear them. Don't forget tape measure. Lock car. Time to go. 10 a.m.

Take path and go around to Steall ruins, or straight up pathless and steep hill to Bealach Cumhann? Short-term pain for two-mile saving? Yes. I go up. Thick grass. Slippy. Urgh. Too steep. Why did I come this way?

Two golden eagles overhead. Rewarded. Stand motionless. Eagles loop above, calling to each other. Effortless soaring on northwest wind. Must get on but can't stop watching. They sense my hurry and disappear over Meall Cumhann. Must keep going. Many miles left.

Reach Bealach Cumhann, 'the narrow pass'. Apt name. Aonach Beag appears ahead. Wears a skirt of white. Not thick. Castor sugar on a cake. Its ancient gaze sees me and laughs across the corrie of the cattle. So far to go. 11 a.m.

Drop into Coire Giubhsachan, 'corry GOOSachan'. Another one. They're everywhere. All over the Highlands. Boggy. Very boggy. Stick to the side of the stream. Grassy and dry. Just jump over the tributaries. Better than walking on bog. Two miles till drier land. Feels longer. Sun comes out briefly. Cold. Think it's snowing on Aonach Beag high above me. Can't tell 'cos of cloud. Winter boots heavy.

Ground steepening. Far less boggy. Sun comes out briefly. Phenomenal scenery. Ben Nevis towering above me on left. Aonach Beag on right. An amphitheatre. The gods are in their seats. Feel like I'm being used as entertainment. Take many pictures. Nobody about. Why aren't more people here?

At last, 830 metres. Final push on to Aonach Mòr. Three-hundred-metre ascent in less than one kilometre. That means 1:3 gradient? Deep breath. Winter boots still heavy. Should've left them. Never mind. Up. Up. Up. Hit snowline. Thicker than it looked from below. Catch breath. Turn around to admire view. Very steep ground. Exposure kicks in again. Hold grass to steady myself. Views across to Carn Mòr Dearg exceptional. Those colours. Snow devil whips across the hill above me. Still no one about. Making good progress. Up.

Reach plateau of Aonach Mòr at last. Cloud lifts. Thick snow. Glad of winter boots. Take it all back. Two walkers on plateau half a mile away. Come from mountain gondola to the north. Cheats. Wait, maybe they walked. Perhaps not cheats. Walk to col of Aonach Beag. Will snow patch be there? Been a while since it's been seen. If so is last one in Scotland. Everything else melted ages ago. How has it lasted? So strange. Please be there.

Reach col. Look down to see if patch is there. Think it is. Can't be sure. Fresh snow obscuring it. Please be there. If gone will be second year in a row. Please be there.

Drop into An Cuil Choire. Steep. Very steep. Loose rock underfoot. Must be careful. No phone signal. No rescue possible if slip. Please be there. Cloud rising. Stags watching me across the corrie. 'What is he doing?' they say in deer language. It's there. I see it. It's big, I think. It's covered in new snow. It's survived. The only old snow in Scotland. Here it is. Tape measure out. Twenty-five metres long. Amazing. So glad it's there. What's the date again? 25 November. Oh yeah. Snow can't melt now. It's too late in year for it to melt.

Stop to have sandwiches. One-thousand-foot cliffs of Aonach Beag above. Utter silence. Unlimited beauty everywhere. Snow patch sits quietly. Sees the same view every day. Sit and work out how it managed to survive. Cliffs? Avalanches? Spindrift? All of those, probably. Never mind. Time to go.

Back the way I came. Hard trudge back on to plateau. Down off Aonach Mòr precarious in snow. Should've gone another way. Rewards worth it. Allt Coire Giubhsachan shining like fire with sun on it. A golden snake slithering for miles. Another eagle overhead. Circles me. Same one as earlier? Has to be. Flies off quickly.

Arrive back at Bealach Cumhann. Feet now wet. Winter boots heavier than ever. Descend. Catch glimpse of car. Not far. View up and down Glen Nevis staggering. Late-afternoon light illuminating silver birch wood opposite. Solitary rowan tree grows out of rock. Steall waterfall behind. Mesmeric. More photos. Too many photos. Tired now. Downwards. Only fifteen minutes to go.

Arrive at car. Last one here. Getting dark. Satisfied at day's work. Last snow in Scotland from last year measured and photographed. Glad.

Winter boots still heavy. Take them off. Ahh.

* * *

26 October 2019, Garbh Choire Mòr

On Saturday 26 October I set out from the Sugarbowl car park at the head of Glenmore, near the Cairngorm ski centre. Heavily laden with a twenty-kilogram pack, my intention was to make for the recently refurbished Garbh Coire refuge, in the shadow of Braeriach. After overnighting there I would, weather permitting, head for Garbh Choire Mòr to see if any old patches of snow had made it from last year. The fresh thick snow that was in evidence on certain aspects meant that if any of the patches from 2018 had persisted to this weekend then it was likely they'd be 'safe' for another year.

Not being in a particular hurry, and wanting to see the England versus New Zealand rugby semi-final, I delayed departure until early afternoon, walking at moderate pace towards the Lairig Ghru via the Chalamain Gap.

The weather was generally fair, but a few squally hail and sleet showers made the going unpleasant at times. Not until I was in the bosom of the Lairig did I have the wind at my heels, which made for an easy enough saunter.

On the way up the Lairig I fortuitously bumped into a couple who had just come from the Garbh Coire refuge, the overnight stop where I was headed. They told me that four Czech walkers were an hour ahead and were likely to get there well before me. Bugger. Given its location, and that it was a cold October day, I'd been sure I'd have it to myself. Not that there's anything wrong with sharing a refuge, of course, but this particular one is very small, and four would be more than cosy. A fifth occupant would have been too crowded for comfort, so I changed to Plan B and instead made for Corrour bothy. This would add a few miles to my journey, but never mind.

It is said that diligence is the mother of good fortune, and so it proved that evening. Upon entering Corrour bothy, some three and a half hours after leaving the car, I came across the Mountain Bothy Association's Neil Reid sitting in front of a roaring fire. He and fellow bothy-fixer-upper Andy had come up to do some maintenance, and we had a good feed and a few drams. A couple of Edinburgh walkers turned up as well, with one of them eschewing his windblown tent for a roll mat on the bothy floor. Wise man.

After a decent sleep and a hearty breakfast, I said my goodbyes and set off back up the path I had traversed the previous evening. About three miles or so battling into a nasty headwind, with freezing rain and sleet thrown in for good measure, I turned off the main path and headed straight for the yawning opening of the mighty Garbh Coire: the Garra-chory. The wade across bog and heather, and leaping the nascent River Dee, was not what you could reasonably call enjoyable. As much as I relish being in the wilds, trudging through boggy ground with a heavy pack is few people's idea of fun. As I drew near to the Garbh Coire refuge, which lay on my path, I could see coloured rucksack covers outside. The snow came on thick now, and the wind howled. The four Czech walkers clearly weren't in any hurry to leave, and I passed the refuge without stopping. Had I knocked on the door I suspect they'd have jumped out of their skins.

The walk up from the refuge to Garbh Choire Mòr, the innermost recess of the Cairngorms, was absolutely foul. Heavy squalls bit into my face, necessitating the use of ski goggles. In addition, the ground was slippery with freshly fallen wet snow, and the goat track that's normally visible in summer was indiscernible. I hopped across bogs and stones, tussocks and holes. Despite decent boots, my feet were getting wet. It was at this point I gave serious consideration to turning back. Looking up to where I was going all I could see was thick mist being blown about by squally snow showers. 'I must be mad,' I said to myself, more than once. But I had come too far to turn back now. I thought about the times I'd been in blizzards so fierce that I could barely even see my own feet. This was nowhere near as bad. Plus, I knew many people were waiting with bated breath to see if any snow patches had survived. Being an avowed mindless optimist, the brief clear(er) spell that arrived gave me hope that the weather would improve. Onwards.

After another half an hour or so's walking, the terrain unyielding in its difficulty, I came to the foot of the Garbh Choire Mòr cliffs. I was greeted upon my arrival by yet another squall. Heavy spindrift danced off the summit plateau and funnelled down the various gullies, giving the corrie an altogether Arctic edge. Despite it being the twenty-fifth time I had visited this location, I had never done so in what was basically full winter condition. It was transformed from its usual benign (if achingly beautiful) state into a place that had no particular liking for *Homo sapiens*.

After telling myself that I was here with a job to do – and an important one at that – I set about it. Depositing my rucksack behind a large stone, I took my small spade to the location that holds the UK's most permanent patch of snow: the Sphinx. Pristine white with freshly fallen snow, there was no discernible old firn, and I feared it was gone.[2] The mild weather of the last two weeks had surely finished it. I stood on the virgin snow and sunk my spade into where I thought any old *névé* would be located.

---

2   *Firn* is from the German for 'of last year', referring to snow that has survived from the previous year.

An odd clunk sounded as the metal hit something about six inches below the surface. I knew instantly that this couldn't be terra firma. I hurriedly cleared away the fluffy white new stuff, and with immense satisfaction saw the dirty remnants of 2018's Sphinx slumbering beneath its new winter duvet. With the weather worsening, I took a few pictures and made a crude measurement attempt (nine metres square). I shovelled as much of the cleared snow as I could back on to the Sphinx to leave it as I found it and returned to my rucksack. Job done. Now for the journey out.

The idea of retracing my steps was not filling me with joy, so I took the decision to go up on to the Braeriach plateau and down to the car via Sròn na Lairig. Though involving more climbing and a risky exit through a steep gully, I figured this would be preferable to the swamp-fest I had ascended by. This seemed like a good idea at first, especially as the snow wasn't very deep on the grassy slope I was going up. When the going got steeper I donned my crampons, giving extra purchase. This was fine at the start, but then the grass gave way to icy snow about thirty metres short of the exit. Looking above I could see quite a cornice had formed. With a twenty-kilogram pack on my back I started to question the wisdom of my exit strategy. I thought about descending, but this would have been riskier than going on. Kicking steps in the increasingly icy and steep gully, I inched up nervously. A slip would have had grave repercussions, and the prospect of being rescued in this seldom-trodden corrie seemed distant. But, eventually, calm and methodical kicking and stepping saw me emerge on to the plateau. My heart was beating quickly from the exertion and the adrenaline. But now there was another fresh challenge.

As I edged on to the plateau I was greeted by a howling gale and bitterly cold temperatures. Allied to that there was virtually no visibility. This wasn't a problem in itself, as I know the Braeriach plateau fairly well (you just follow the rim of the corrie, using it as a handrail). Walking into the wind was energy-sapping, particularly across a seemingly never-ending boulder field. No matter how many times I thought, 'The summit's just over the next rise', there was inevitably another one. When the summit eventually did come into view, I was no more than ten metres away from it, such was the poor visibility. From there it was a long, long descent down Sròn na Lairig and back through the Chalamain Gap and to the car. Fifteen hard miles.

With all that said, I drove back home with an incredible sense of satisfaction and achievement, knowing that at least one of Scotland's tiny relics from last winter had lived on to fight another year.

* * *

26 September 2015, Sgùrr na Lapaich

I was awoken at 5.30 a.m. by my alarm. It was going to be a long day, so an early start was essential. Rousing myself from the warmth of the sleeping bag, I peered out of the tent. It was dry, thankfully, and the clouds sufficiently few as to allow the first metallic blue streaks of the approaching sunrise to announce themselves. Two stags already roared on the hillside above me. It was, after all, the eve of the rutting season. I took a quick breakfast but decided not to pack the tent because of the amount of dew on the outer sheet. The forecast was dry so it could wait until I came back later in the day.

I set off with heavy legs along the northern shore of Loch Mullardoch. The sun was taking its time to rise, so I had to be careful picking my way through the bracken and stones. This was not a place to have a mishap. It is the very wildest of country, far from help and phone reception. The more sensible walkers, when picking off the four Munros on the north side of the loch, opt to hire the boatman who ferries people to the far end of Mullardoch. I had no such luxury. The walk to the base of An Riabhachan took over two and a half hours, and by the end had turned into a penance. I was happy, therefore, to leave the unceasing rock and bracken behind as I started to gain height.

This was the first time I had set foot in these fine hills. Loch Mullardoch and its northern neighbour Loch Monar occupy some of the most isolated countryside the Highlands have to offer. Long had I promised myself I would visit them, but always found a reason not to. This autumn there really was no option. So much snow still lay on the hills from the previous winter that a visit there was essential to gauge how much was likely to survive until the following winter. Such was the remoteness and roughness of the terrain that I could not rely on waiting for someone else to do it.

The walk up An Riabhachan's south-east shoulder, Sròn na Frìthe, was a steep but enjoyable one. The ground was dry and the view across the loch to the high hills opposite was glorious in the now first proper light of the day. Coming this way, where so few had ventured before, meant that there were no paths to make the ascent easier. This hardly mattered to me. My approach to climbing hills is, and has always been, 'If in doubt, keep going up.' (If one adopts this principle in general, the only logical end point is the summit cairn.)

Before long, the summit ridge had been attained and from there it was an easy saunter to the summit of this majestic hill. On the cliffs of its north-east corrie, Creagan Toll an Lochain, lay a huge wreath of snow. In excess of 100 metres long, it was certain to survive until next winter – for the first time in twenty years. I took some photographs and had a sandwich. The next hill to be climbed was Sgùrr na Lapaich. There were rumours of large patches still on this hill. I was keen to see them.

Duly climbed, I dropped off the summit of Sgùrr na Lapaich. It is the highest point on the long range of hills which rise above Loch Mullardoch's northern flank, and the most handsome. Its shapely ridges glide down gracefully and evenly on three sides. As I eased down the middle one, which runs due east, I became aware of a huge scar on the headwall of the hill's unnamed south-east corrie. The farther down the shoulder towards the Bealach na Cloiche Duibhe – 'the mountain pass of the black stones' – I descended, the more of the corrie I could see.

'That can't be snow. No way,' I said to myself. 'God almighty! It is.' I stood rooted to the spot, my brain quite literally unable to process the vision that was now before me. I was at this point about 300 feet above the corrie floor, still high on the ridge above. Down on that corrie floor was a massive bank of what appeared at first glance to be snow. But it wasn't ordinary snow. It was almost entirely covered in a brown, muddy mixture of earth, rock and water. A small stream had gouged a channel all the way through it, though this itself was partially hidden by the sheer volume of material surrounding it. I had to get closer. Down I traversed through steep grass and rock for the next ten minutes, until at last I was as close to the snow as I dared get.

It was a scene of utter devastation. Nowhere was there any sign of ground level or vegetation. The floor's expanse in front of me looked like the beginnings of a large construction site. Mud and large rocks lay everywhere. Some of these rocks were huge, many tonnes in weight. Even more strange-looking than this, however, was what this debris sat on. At the edges of the rubble and wreckage, deep snow could quite clearly be seen. It was completely discoloured by the mud that was leaching into it. It looked, to all intents and purposes, like the snout of a glacier. It looked old. So, what had caused this ruin? It could be only one thing: a massive avalanche.

As mentioned previously in this chapter, 2015 was an unusually snowy year. Huge accumulations of snow had built up on some aspects all the way through April and even May. As a result, there were colossal, droop-ing cornices across the Highlands in June just waiting to fall. And fall they did, nowhere bigger than here. I was to find out later in September that, amazingly, there was a witness to this avalanche.

On 20 June 2015 a walker was out rambling on Sgùrr na Lapaich in thick mist. Soon after summiting she heard 'an almighty rumbling and crashing as the snow slope on the eastern side of the coire detached. We were above it but saw the debris later.' It is incredibly fortunate to get an exact fix on this date, not least because it shows that big avalanches can occur in Scotland even at midsummer: literally. Eight days later another walker passed this way and took photographs. What they show is remarkable. Thousands upon thousands of tonnes of snow, rock and earth had slid right from the top of the hill and smashed aside everything in their way.

As I stood that day on 26 September, three months after the event, the layers were still very much in evidence. The profile of the snow was fascinating. It was like a birthday cake with two very distinct layers. On the top, perhaps two feet or so thick, was a disfigured and mangled layer, interspersed with rock and mud. This was clearly the avalanche layer which had slid down. It was lying on top of the winter snow layer, which was far more uniform and complete. It was also far deeper, maybe five or six feet in places. Capping both these layers was the mud, earth and rock. Oddly, despite the chaotic scene, the top layer of earth and mud was giving the snow beneath a protective layer, hence preserving it. In other words, had the avalanche not happened then all snow here would long ago have melted. As it happens, this particular snow survived the year, where it had not done for many before.

The scene remains the most memorable I have ever encountered in Scotland, and the most unusual. To see so much material dislodged was slightly disconcerting, as was the volume of snow that it preserved for the year. It was yet another experience that, had it not been for my hobby, I would never have had.

Top The view from the author's parents' house across to Ben Lomond, the hill which was the catalyst for the author's passion.

Above Shafts of sunlight illuminate a glen near Aonach Beag.

Top Ecologist, mountaineer, polymath. The late Adam Watson conducting the annual snow-patch survey from Meall Odhar near Glenshee ski centre in July 2009. © *Allan Cameron*.

Above Britain's snowiest place, Garbh Choire Mòr on Braeriach, taken during the snowiest winter of the twentieth century, 1951. Exceptional depths here, perhaps up to seventy-five feet deep. © *Adam Watson*.

Top The place of the eternal snows: Braeriach's Garbh Choire Mòr in typical late summer condition.

Above The author at Britain's longest-lasting patch, the Sphinx, at Garbh Choire Mòr. The snow was to survive that year, but by the time it was buried it measured the size of a small dining table. © *Murdo MacLeod.*

The towering cliffs of Garbh Choire Mòr on 11 October 2019. The patch in the foreground, Pinnacles, is named after the granite tors above and melted completely just a few days later. It is Scotland's second longest-lasting patch of snow historically.

Top Britain's longest-lasting snow patch, the Sphinx, a day or so away from melting on 30 September 2017. The snow in the author's hands fell in late 2006 and was almost eleven years old.

Above The author at the site of the Sphinx. The ground here had been exposed to the elements for perhaps only a few months in the last 300 years when visited by the author in late September 2018. This was the second year in succession that it had melted.

A walker is dwarfed by the huge remains of winter snow still present in the Cairngorms in early September 2015.

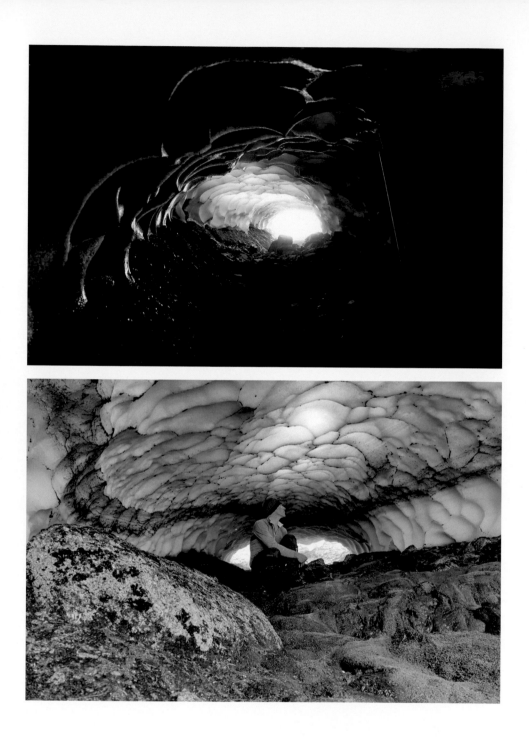

Top Only certain colours of the light spectrum can penetrate snow, amply shown here in a magnificent snow tunnel near Drumochter in the Cairngorms National Park.

Above The author sits under a remarkable snow tunnel at Beinn Bhrotain in the Cairngorms. The range of colours displayed both on the snow and the ground were exceptional.

six

Dr Adam Watson

Exploring for yourself by your own free will, without formal courses,
training or detailed planning, is the best joy the hills can give.
— Dr Adam Watson[1]

With the honourable exception of my parents, no one has had a bigger influence on my thinking, and nobody at all with regard to my approach to the outdoors, than Adam Watson. In the period I knew him, from 2005 to his death in 2019, I learnt more about research, writing, snow and the general outdoors than I did in the previous thirty-odd years.

It is almost impossible to overstate the importance of Dr Adam Watson to the understanding of Scotland's upland ecology. Knowing where to begin when listing his achievements, awards, testimonials, publications, etc. is almost a futile task, and would need its own chapter.

Adam was known universally in Scottish scientific and outdoors circles. Whether it was birds, place names, flora, fauna, it didn't matter: if he said something, it automatically commanded respect, with people drawn to him because of his unparalleled knowledge. Widely acknowledged in

---

1   Watson, Adam (2011), *It's a Fine Day for the Hill*, Bath: Paragon Publishing.

his time, even amongst chiselled mountaineers, as *the* pre-eminent expert on the Cairngorms, he walked, skied and climbed on that range for the best part of seventy-five years. His wide array of studies and associated published output, not just on the Cairngorms but across Scotland and as far as Arctic Canada, almost beggar belief.

As if all this weren't enough, he has, single-handedly, done more on the systematic study of snow and snow patches in Scotland over the last fifty years than virtually all other researchers combined. Moreover, to be frank, were it not for him the chances of this book being written would be small. Remote, even.

There is room only in this chapter to touch on some of his output outside of snow research. It would be remiss, though, not to recall just a few of his general writings and the backstory of this most remarkable of men.

* * *

Adam Watson FRSE, FRSB, FINA, FRMS, FCEH, was born on 13 April 1930, in Turriff, Aberdeenshire.[2] He grew up in the town, where his father was a solicitor and had his own practice. Despite a serious childhood illness which kept him from school for some time, Adam excelled academically. He was awarded the *dux* prize in secondary school and went on to attend Aberdeen University, where he studied zoology and graduated with a first-class honours degree in 1952.[3]

As a young boy, Adam developed a fascination for a multitude of subjects. A keen birdwatcher from an early age, he would frequently note nesting sites and behaviours in his diary. So precise were his notes that when a respected local birdwatcher turned up at his father's practice to speak to an 'Adam Watson' he was amazed when the teenage Adam (not the forty-eight-year-old father, also called Adam) appeared and started chatting to him.

---

2   Fellow of the Royal Society of Edinburgh, Fellow of the Royal Society of Biology, Fellow of the Arctic Institute of North America, Fellow of the Royal Meteorological Society, Fellow of the Centre for Ecology and Hydrology.

3   The *dux* prize is a Scottish academic award, usually given to the highest-performing pupil in a school in a given year.

Adam's epiphany on snow, the subject which it is fair to say fascinated him more than any other in life, happened in 1937 when he was just seven years old. He later wrote:

> I looked from a window as pale columns of snow crossed a dark sky and big flakes tumbled. For hours I watched flakes melt on a slate roof and then gradually persist until the snow hid all the slates. In the garden I waded in the deep snow, feeling it cold in my fingers until it turned grey and melted. A foot of white became an inch of grey as I stood on it. Each snowflake had a variety, and each of the crystals comprising a flake. The airy mantle on branch, pavement and road absorbed sound. A strange hush pervaded the normally noisy town.[4]

This was his spark. From being fascinated by snow in general, his attention soon turned to observing snow patches in summer. By the age of nine he had already drawn the shapes of some of them in his diary and notepads. Mirroring my own interest as a nine-year-old, he was curious as to why they lasted so long.

It was also at this age when Adam picked up a book. A small, blue book. On a wet July day in 1939, during a family holiday on Deeside, he happened across a copy of Seton Gordon's seminal *The Cairngorm Hills of Scotland*. It literally changed the young Adam's life:

> Books are one of the pinnacles of human culture and achievement. Few things can have a more revolutionary effect on the attitudes and beliefs of young minds in a receptive mood. That was so for me, with Seton Gordon's books. Only perhaps once or twice in a lifetime may a brief event, such as a casual glance at a book, or a sudden union of two like minds, become a clear turning point which transforms the rest of one's life.[5]

4   Watson, Adam (2011), *It's a Fine Day for the Hill*, Bath: Paragon Publishing.
5   Gordon, Seton (1979), *The Immortal Isles*, Lewiston: Melven Press (from the foreword by Adam Watson).

Contained in Gordon's book were notes and photographs on a host of topics, and all of them struck a chord with the nine-year-old Adam. He studied each and every page with wide-eyed interest. One topic, however, caught the young boy's eye more any other: snow. Adam was excited, as I was years later, to learn that other people were looking at patches of snow on remote Scottish hills and recording them. A kindred spirit, he thought!

There was nothing else for it: he decided after reading the book that he *must* write to Seton Gordon to share some information with him. To use his own words, he had a 'burning desire to write to the man who changed my life'. So, on 10 October 1939 he did just that. Expecting no response, Adam was beyond excited when Gordon wrote back a week later. Thus commenced a correspondence that would last for the next thirty-eight years.

Along with snow, another of Adam's great passions was the ptarmigan. This wonderfully adapted member of the grouse family can, in the UK, only be found in the Highlands of Scotland, and even then generally above 3,000 feet. It is the only bird in the UK whose plumage changes in winter to brilliant white. As well as being the topic of Adam's degree dissertation, the ptarmigan was the focus of his PhD study and thesis, which was achieved in 1956. All through his life he would return to the ptarmigan, and he wrote on it extensively – becoming one of the world's foremost authorities. Though I never asked him about why he found the ptarmigan – specifically – so interesting, I suspect that part of his great love for the bird was that he envied it. Often he spoke of how we, humans, were but a fleeting visitor to the upland environment. Though we may enjoy a day on the highest tops, we are always obliged to return to our homes either by hunger, a need for shelter or to escape harsh weather. The ptarmigan need not return to low ground for any of these things. To sate its hunger, it eats plants which grow on the rough ground that covers the highest hills. For shelter, and to escape the worst of the weather, it can fashion a snow hole in a matter of seconds. From there it sees out the worst that the weather can throw at it, in relative comfort. The bird also, Adam stated more than once, taught him humility.

A diary entry from 10 April 1944, when he was just thirteen years old, recalls his first encounter with the bird. The quality of the writing and the eye for detail is astonishing for such a young boy:

> At the top of the shoulder I had lunch above a big snowdrift, and passed more drifts further up. The summit has many boulders and a cairn at 2,860 feet. Mr Seton Gordon told me that if I climbed Morven I might see ptarmigan at the top. Not having seen any, I determined to try and find one before stopping to look at the view. A moment after, when I had gone ten yards down the north side, a beautiful cock ptarmigan with a fine crimson comb suddenly rose from my feet and ran down the hill after a short distance. When I followed, it ran further, and then flew to alight at 2,500 feet. Returning to the cairn, I looked at the view, simply magnificent. Loch Muick lay in a hollow, and I saw Lochnagar ... sharply set against a slightly hazy sky ... A spotless snow blanket covered the rounded top of Ben Macdui, while through the telescope I saw the big black cairn standing out sharply against the surrounding white. Ben Avon looked as though whitewash had been emptied over it.[6]

His precociousness and supreme self-confidence were in evidence all through his adolescence and into early adulthood. Never, though, was this self-assurance worn with any hint of arrogance. It came out of a profound and deep knowledge of the upland environment and was attained and maintained by operating extensively within it from a very early age. His confidence manifested itself in some of the extraordinary adventures he undertook. None exemplify this better than his trip to Iceland in 1949, not long after his nineteenth birthday. Armed with only ten pounds in his pocket, he and two university friends 'hitched' on an Icelandic trawler that had put ashore at Aberdeen for a brief time. Gathering their gear after a botany exam at Aberdeen University, they

---

6   Watson, Adam (2012), *Some Days from a Hill Diary: Scotland, Iceland, Norway, 1943–50*, Bath: Paragon Publishing.

rushed to the harbour and boarded, not really knowing what to expect when they arrived. All that mattered was the adventure that awaited them. When they reached their destination, those adventures began, and in earnest. Skiing and climbing among the snowy peaks were the order of the day, with nature-spotting thrown in during the more relaxing periods. Being short of cash, they had to improvise with foodstuffs. The country was very expensive due to inflation, so they'd often flush wildfowl off the nest and fry or boil the eggs that the birds sat on. The three young men would have salted cod during most meals, which they'd bought a great store of for not very much money, due to its great abundance in Icelandic waters. Reading about the escapades of skiing off great snowy peaks one day, then climbing cliff faces the next, it is obvious the trio enjoyed a memorable time in the country. Adam enjoyed it so much he stayed on an extra week himself.

Before leaving Iceland, after a month's adventuring, Adam was put up in a police cell whilst waiting for a trawler to take him back to Scotland. There was no crime in the area, so the prison was empty. Adam was very grateful to the local policemen for the bed in this most unusual of digs.

The exploits of the three friends are chronicled in several of Adam's books, and they leave readers shaking their heads at the audaciousness of it all. Trying to imagine three nineteen-year-olds undertaking the same thing today is very difficult. The red tape alone would be enough to deter the most resolute and intrepid students.

* * *

By the time Adam was nineteen and had been to Iceland he was already a good friend of Tom Weir, that most well-known of Scottish outdoors' characters. They met near Braemar, at the home of Cairngorms stalwart Bob Scott, in December 1947. Tommy Weir, as he was known to his friends at the time, introduced Adam to ski mountaineering: a skill that would transform his enjoyment of the hills and his ability to get around them. Weir influenced Adam's attitudes to hillwalking and mountaineering more than any other person he'd ever known, Seton Gordon included. Often in Adam's writings he references the role Weir played in his understanding of what was possible when on the hill. He found Weir's enthusiasm most infectious. A fine example that he wrote about was from March 1949, when Adam and Weir were skiing in 'very poor conditions' on Braeriach, but with the hope that the weather would improve. Sure enough, it did:

> As we reached the summit of Braeriach, we suddenly saw blue sky and immediately the sun began to turn firm in the frosty air. Just as we began to see extraordinary views of snowy slopes and cliffs leaping out of clouds, ... Tommy shouted 'Great stuff!' and on hearing this I felt like leaping into the air with excitement.[7]

Tom Weir and Adam enjoyed many days out on the hill together. Adam's chapter on Weir in his book *It's a Fine Day for the Hill* goes into some of them in detail. It shows the enormous respect Adam had for Weir, as well as some of the latter's less attractive sides. As is often the case with great people, their lack of ability sometimes to compromise or see the other's point of view can be to their detriment. That certainly seems to be the case with Weir.

But of all the foreign expeditions Adam made in his teens and twenties, I got the distinct impression from him, in person and via his writings, that the one he held most dear was the trip to Baffin Island, Canada, in 1953. He spent four months there in summer as part of an international expedition by the Arctic Institute of North America. His role in the group was to study birds and mammals. It is little exaggeration to say that I lost count of the amount of times that Adam mentioned this trip. The accompanying book that he wrote about the journey, *A Zoologist on Baffin Island, 1953*, is, to my mind, one of the finest things that he wrote. None of the writings are more touching, or better evoke the sense of joy that the trip gave him, than the last paragraph of the journal:

---

7   Watson, Adam (2011), *It's a Fine Day for the Hill*, Bath: Paragon Publishing.

On the day I left … [I was gifted a] copy of the Swiss song-book *Poly Liederbuch*, which contained most of the songs that we sang on the expedition. By a nice coincidence it was published in 1953. Later I wrote on it 'From Hans Röthlisberger 1953'. During 1981, when Fritz called at my house on his way to give expert advice at a Public Inquiry, he and I sang *Aprite le porte*. Then he added his signature 'Franz Hans Schwarzenbach 1981'. I treasure my wee green book. Now as I open the pages fifty-eight years later and see the favourite songs of our Swiss team, my eyes fill with tears.[8]

When one reads the book that contains the accounts of the Baffin trip it is obvious to see the seriousness with which the team took the expedition. Belonging to an era where, academically, rigour was paramount, Adam believed passionately in testing his theories in the real world to see if they stood up. If they didn't, he would discard them. As he saw it, scientists who were uncritical in dismissing the possibility of an alternative hypothesis were doing no one any favours. He very much disliked it when academics were so precious about their work that they recoiled from proper scrutiny. For Adam, the truth was far more important than hurt feelings. How we need such thinking today.

Time and time again he spoke with passion on the value of fieldwork. He lived and breathed for getting out on the hill and gathering evidence. So much so, in fact, that by his own admission it was sometimes difficult to tell where work stopped and free time began. 'I just liked working,' he would say.

There are simply too many other stories and people for this book to go into when discussing Adam's past. Luckily he left a large body of work for us to enjoy and peruse. It is seldom far from reach on my bookshelves.

* * *

Adam's work on the subject of snow and snow patches is astonishing. From the age of eight he would note and record individual patches.

---

8   Watson, Adam (2011), *A Zoologist on Baffin Island, 1953*, Bath: Paragon Publishing.

He continued this throughout adolescence and adulthood, right up until his very last days. In the many thousands of emails, telephone calls and physical meetings I had with him from 2005 to 2019, he never tired of talking about them. He would often say to me on the phone that he had been out on a drive on Deeside with his friend Derek Pyper, and that they had pulled into some lay-by or other to observe the patches on Lochnagar, or Beinn a' Bhuird, or Brown Cow Hill, and so on. Every trip in good weather, to count hare, grouse, ptarmigan or whatever, he would also mark the patches in his little notepad. Not only did he do it in the name of research, but for the sheer love and joy of it.

The annual 1 July survey from Meall Odhar near Glenshee ski centre, which is discussed at length in the next chapter, was and is, for me, one of the highlights of the year. So important did I find doing it with Adam that in 2010 it got me into trouble.

On 25 June 2010, I married my fiancée in a church in Essex. I lived in Chelmsford – Essex's county town – at the time, and my fiancée was from that part of the world. During the preparations for the wedding, the subject of the honeymoon raised its head. In truth I was not looking forward to this conversation, even though I knew it was something I could not avoid. It is established tradition that newlyweds honeymoon very soon after a wedding, perhaps a couple of days thereafter. This tradition did not suit the timing of the 1 July survey, however. I did not want to miss it, so I decided that attack was the best form of defence. I boldly announced one evening that any honeymoon would be impossible before at least 3 July that year. The annual snow survey *must* take priority. The *only* way around this would be for us to honeymoon in Scotland. (I knew there was little to no chance of this happening, so it was a calculated risk with very long odds.) My fiancée looked surprised when I announced this, but I was resolute, and she accepted this short delay. In the intervening years I have told this story to various people, to varying levels of astonishment. I am normally a very easy-going person and don't rock boats any more than I need or have to. In this, though, she could see I was resolved, and she did not object in the end, seeing how much it meant to me. In the event we departed for Morocco well after survey day, much to Adam's amusement. (The story about how I absconded from the honeymoon for a day to go on a trip up Jbel Toubkal, North Africa's highest mountain, to look at Moroccan snow patches, is for another day.)

The 1 July survey was also the first time I met Adam, in 2008. I remember it well. We were to meet at Glenshee ski centre, whereupon the then centre manager, Graham McCabe, would drive us up in his Land Rover to the summit of Meall Odhar. It was a lovely morning as my car pulled into the car park that day. The air was clear and the cloud base high: perfect for snow-spotting. Soon after I arrived, a small blue/silver vehicle pulled into the car park and I instantly recognised Adam by his long, flowing white beard. I introduced myself to him and was met with a large, beaming smile. On that day I spent many hours in his company, learning much even in that time. I did not miss a subsequent July survey in all the time Adam was alive. In fact, in 2018, when he was unwell, I was charged with undertaking it myself. Determined, I went on to the Cairngorms immediately after doing the Glenshee survey, thereby completing an additional ten-mile round-trip to conclude the survey. In his reply to me the next day he called it a 'tour de force'. I was extremely pleased and proud at this.

The memories of each 1 July survey are precious to me. Though in reality we endured a lot of different weather over the years as we did it, in my mind all I can see is us sitting on the warm grass in brilliant sunshine, each with binoculars, counting every patch together. Latterly, either because he wasn't sure himself or he trusted my judgment an increasing amount, he would ask me how big I thought a particular patch was.

'Is that a four or a five, would you say, Iain?'

'Looks like a five to me, Adam.'

'Aye, I think that's right enough.'

The 'four' or 'five' he talked about was a numbering system he had devised for measuring patches from great distance. It is very difficult to judge whether something is ten metres or fifteen metres long when one is several miles away. It is much easier, though, to estimate if its size is between five and ten metres, or ten and twenty metres. Adam had a system for everything. I still use it now.

The 2018 survey was the first one that Adam had missed since he started doing it in 1974. It was a worry. For a number of years, we had seen that he was becoming increasingly frail. He walked with a stick constantly by then and was unsteady on his feet. Despite this physical frailty, however, his mind remained razor-sharp. At a gathering of friends some time before the 2018 survey he regaled the assembled crowd with stories on place names, folk tales and snow patches. He never missed a beat with his timing.

At his eighty-sixth birthday party, in Braemar, many of Adam's friends and acquaintances gathered to celebrate. It was quite a large group and some of the attendees did not know each other, or at least had never met. Adam did a round-the-table, introducing each person and giving a brief synopsis of who they were. There were some highly impressive people present, with some renowned authorities in their fields, and as some of them were announced I suddenly felt quite inadequate. With so many academics and experts, I lowered my head when he got to me. In his characteristically generous fashion, Adam introduced me as 'probably *the* expert on snow patches in Scotland'. I was taken aback. Immediately I replied, 'Oh, I don't think so!' or words similar. Without pausing he replied, 'Oh, I think so, Iain. Nobody has put more work in than you have.' Such generosity and sincerity one does not easily forget. What, I think, he was doing when he said this was figuratively passing on the baton. He was happy that I had earned my stripes and was ready to take on the task that he'd been doing for almost eighty years.

Though there are many people fascinated by snow and – more recently – snow patches, the amount of folk who are willing to put their boots on and go on to the hill to measure, record and photograph them is still relatively small. But there are innumerably more people now than when Adam started doing it in the 1940s. This is what makes his archive of information so impressive. Even today, with modern technology and the availability of information, our records cannot compete with his. He listed literally thousands of patches across the Highlands and was able to monitor a lot of these when out doing other fieldwork. I am still working through an archive of his files and am consistently amazed at the depth and sheer amount of information that is contained within. It seems bottomless.

* * *

On the morning of 23 January 2019 my phone rang. It was Adam's daughter, Jenny. I knew that Adam had been ill and was in hospital, as I had spoken with Jenny just a few days before. She had been concerned as he wasn't eating or drinking much. Immediately I knew what the news was going to be. I answered the call with an already shaking hand.

'Oh, Iain. Dad died this morning,' said Jenny, her voice cracking. My heart sank. I confess that at that point my mind went blank and I remember little of the conversation. It was a gut-wrenching piece of news to hear. I think I tried to say something comforting and consoling, but I probably didn't. Her grief was all too evident down the line.

I do not profess to have known Adam nearly as well as many others. To them his loss will be more keenly felt than for me. But, on that morning in January, it felt like a little piece of the soul of Scotland had slipped away. Hyperbole is all too easy when talking about great people, but I think in Adam's case every bit of it is deserved. His work, completed over seventy-five years of being among the birds, hills, lochs, rivers, people, trees and snow, has left an indelible imprint on the collective understanding of the country that he loved so dearly.

Adam taught me the value of discipline, rigour, patience and diligence when it came to record-keeping and the general gathering of information. As befitted his academic background and training, he also cautioned against overexaggeration – something that I must admit I have a natural preponderance towards. I can see his face in my mind if ever now I think about plumping up a story.

Adam's childlike wonder and awe for the works of Mother Nature were attributes that never left him. Whether he was watching snow clouds form over the North Sea, or inspecting a grouse walking across a mountain path to ward off humans from its chicks, he always had that glint in his eye. His deep, deep passion for the natural environment – especially his beloved Cairngorms – was infectious. Though my own passion for what I do is significant, it has been enhanced by knowing Adam. I have never met anyone like him, and I don't suppose I shall ever again. He was, truly, a one-off.

The very simplest and best writing I can think of to sum Adam up is that which is contained in Seton Gordon's first reply to Adam's letter to him in October 1939. Never has there been a finer, more prophetic or more apt summation of such an important man written in so few words:

It is a fine thing for you to have a love of the hills, because on the hills you find yourself near grand and beautiful things. And, as you grow older, you will love them more and more.

I last saw Adam in December 2018, just a month before his passing. By then he was extremely frail and had been diagnosed with a terminal condition, but he managed a scone and a small bowl of soup at his favourite café, near his house on Deeside. With us was another of his friends, Mick Marquiss – himself a renowned and highly respected bird-watcher. We chatted for a couple of hours about digitising Adam's snow-patch observations. He was full of ideas and chat. He also had the same beaming smile and glint in his eye as when I met him all those years before at Glenshee. As I left that day I wondered if I would see him again. I am glad that our last meeting was a happy one.

* * *

In April 2019 I was asked to write an obituary for Adam. It appeared in the Mountaineering Scotland magazine shortly afterwards. It is tran-scribed in its entirety below. Reading it for the first time in a while, I smile as I remember some of the memories:

A number of years ago, while we were conducting the annual July 1st snow survey at Glenshee ski centre, I asked Adam what his favourite bird call was. With his almost unparalleled experience of fauna in upland Scotland and similar arctic environments, I expected something a bit more exotic than the answer he gave. Perhaps the snowy owl, I thought, or the metallic drum of the snipe? Maybe even the evocative call of the peewit as it swoops over the Aberdeenshire farmland?

'I like the ptarmigan,' was his reply.

No one who knew Adam will be surprised by this answer. I shouldn't have been. But it so was typical of the man. Understated and considered, his reply was – in many ways – a mirror image of his character.

I first contacted Adam in 2005, after having read a piece he had written about snow patches in Glen Coe. His email address was at the bottom of the text, and I thought he might be interested to learn of my diary entry for 1993, which ran contrary to the information he had presented. I was, of course, wary of doing so. He was a well-known Scottish scientist of immense standing and respect. Here was I, a nobody, who had left school at sixteen and became an electrician, without an academic bone in my body. However, as I came to understand, our interest in patches of snow that clung on to the shady recesses of Scotland's highest hills was something that united us in fascination, bemusement and wonder. It transpired that he, like me, became hooked by this esoteric area of study as a child. Anyway, I needn't have worried. In his effusive reply he thanked me for my note and asked if I had any observations for that year, and any others that I might have. I was thrilled! He was apparently completely unconcerned by my lack of academic background. 'Enthusiasm is the most important characteristic a person can have in the field of study, Iain,' he told me. It was a word I heard him use many times.

Over the course of a normal person's life you might meet a few people who you really rate. People, for whatever reason, whose knowledge seems almost limitless, or whose grasp of a subject, or subjects, is boundless. Years – decades – of study and immersion have generally been spent in pursuit of this knowledge. But that will only be part of it. When they speak to you about their subjects they generally do so with an accompanying passion and elucidation that few others can.

Folk like these are rare. Adam was one. In the field of outdoor studies on the high Scottish landscape he was virtually peerless, especially in his beloved Cairngorms. It is doubtful that any one person has known as much as he did on the flora, fauna, people, place names, and ecology of this range, and nor is there ever likely to be one in the future. He spent his formative years in the Cairngorms, and the list of people he met, befriended and spent time with there during the course of his life, whether walking, climbing or skiing, reads like a *Who's Who* of the great and the good of the Scottish outdoor world. Folk like Tom Weir, Bob Scott, Tom Patey, Mac Smith, Seton Gordon. The list goes on.

As much as spending time in Adam's company was informative, it was also good fun. He had a childlike quality of mischief in some of his dealings. Well do I remember sitting having lunch with him and his beloved wife, Jenny (who died in 2016), and being regaled with possibly the best story I had ever heard. (The tale involved a dispute between a member of the royal family and a local, with the latter telling the former to remove himself from his land, using very Anglo-Saxon language. But that's a story best not committed to print.) Another episode was during the winter of 2010 when he and I went to Arbuthnott kirk near Stonehaven to see the grave of Lewis Grassic Gibbon, author of the classic novel *Sunset Song*. The graveyard was like an ice rink. Literally. Water had poured down an adjacent embankment on to the grass and had frozen solid. We entered the church grounds and had to hold each other up, laughing as we slid over to where Grassic Gibbon was buried. Adam commented that it would be ironic if we were to slip and meet our ends in a graveyard.

One of Adam's biggest assets, though, was his generosity of spirit and his patience. This generosity was very evident in 2011, when he asked me to become lead author of the annual snow-patch paper that he had written every year since 1995, and which appeared in the Royal Meteorological Society's *Weather* journal. Perhaps aware that he wasn't going to be able to do it forever, he suggested I had a bash. I recall my first attempt. It was returned with so much red pen that I thought maybe I wasn't cut out for it. Adam wouldn't hear any of this and encouraged me with what words to use (and, crucially, what words I had used too often, or incorrectly). It makes me immensely sad now to think that I shall never get another opportunity to send Adam the draft for his comment.

Adam Watson was an inspiration. He will be mourned by his family, and by the Scottish scientific community. All lovers of the Scottish outdoors are in his debt. I shall miss him.

seven

Scottish snow and climate change
(with a eulogy to winter)

I t is no exaggeration to say that the very longest-lasting patches of snow in Scotland exist right on the edges of being able to survive from year to year. Whereas in previous decades it could be said with confidence that year-round snow was the norm, it is increasingly the case that this mindset is an artefact of history. How dated the outlook of the Scottish Mountaineering Club members who wrote to *The Times* in the 1930s looks to us now, when they pronounced that none of them expected snow ever to vanish again from Garbh Choire Mòr. In the modern era any notion that our snow could be permanent is misplaced; the patches at Garbh Choire Mòr are walking a tightrope without a pole.

Even the language we use to describe these snows is at risk of becoming obsolete. Once upon a time the adjective that best described them was *perpetual*; later it became *semi-permanent*, and then *semi-perennial*. But this phrase, too, will not be able to withstand the pressure put on it by the snow's continuing disappearance. Does something that vanished entirely in 2017 and 2018, and only just made it through the year in 2019, deserve the epithet of *semi-perennial*? This may seem a semantic consideration of little consequence in the real world but, on a planet where epochal changes have – in the past – been measured in millennia, the relatively rapid realignment of descriptive wording

» From 1900 to 1980 a new temperature record was set on average every 13.5 years. From 1981 to 2019, it increased to every three years.[3]

These figures are very persuasive. And, although it is extremely difficult – not to mention potentially unwise – to extrapolate one set of data and link it to local variations elsewhere on the planet, as a general trend it is relevant to introduce evidence of a warming planet into any discussions about a decrease in snow cover generally.

* * *

The snow's demise

Since 1974, an annual Scottish snow survey has been done on 1 July. This survey is conducted each year from the same place: Meall Odhar, a hill overlooking the Glenshee ski area in the North-East Highlands. Sitting just above 3,000 feet, its summit is one of the finest vantage points in Scotland for the number of hills one can see that still carry snow. The object of the study is to count *every single* extant patch of snow lying on hills east of the A9 road. It is an ambitious thing to undertake every year. Not only because of the sheer scale of the survey area, but also because of the weather. Often 1 July is a washout, so a day or two either side needs to be considered if the forecast is poor. Likewise, if the number of snow patches visible on this date is huge – as it so often has been – then extra eyes, and accompanying legs, are drafted in to assist. Each patch is recorded visually because this is the only practicable way of counting (measuring each patch would be impossible unless we could draw on literally hundreds of volunteers). When seen from the Meall Odhar viewpoint, the largest snow wreath is almost always on Ben Macdui, and is often as much as a kilometre long. Others can be less than five metres. If one splits the survey totals since 1974 into two sixteen-year chunks (i.e. the first sixteen years, 1974–1989, and the last sixteen years, 2004–2019) then one can see the results. They are sobering.

---

3 www.climate.gov/news-features/understanding-climate/climate-change-global-temperature [accessed 13 May 2021].

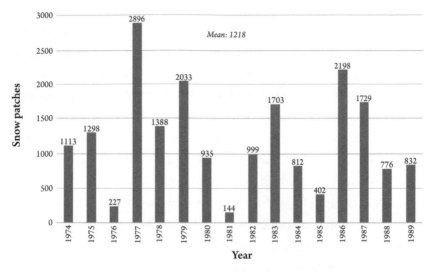

*Figure 1: Total number of individual snow patches counted
in the North-East Highlands on 1 July (1974–1989)*

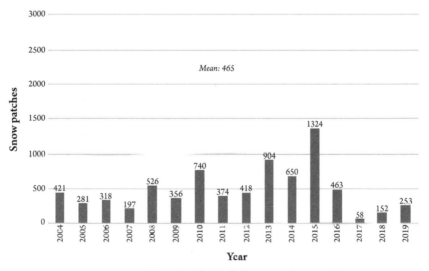

*Figure 2: Total number of individual snow patches counted
in the North-East Highlands on 1 July (2004–2019)*

The mean (average) total of patches has more than halved from 1,218 in the period 1974–1989 to 465 in 2004–2019. This is a quite remarkable reduction in a relatively short space of time. During the first sixteen-year range, totals of 1,000 or more were commonplace, and in 1977 getting on for triple that amount. In the most recent period only once did it exceed 1,000 (in 2015). If one plots a trendline on a graph of these totals then it is obvious to see which way it is going: down. It also seems to suggest a correlation with the NASA figures, which show the ten warmest years on record have been some of the most recent.

As we may recall from chapter 4, the number of times that all snow in Scotland vanished has increased markedly since 1996 (four times between 1996 and 2019, compared to just twice in the previous 200 years). This on its own ought to be a cause for concern. If we look slightly beyond this, though, and examine the times that all snow *nearly* disappeared between 1996 and 2019 then the picture becomes even more stark.

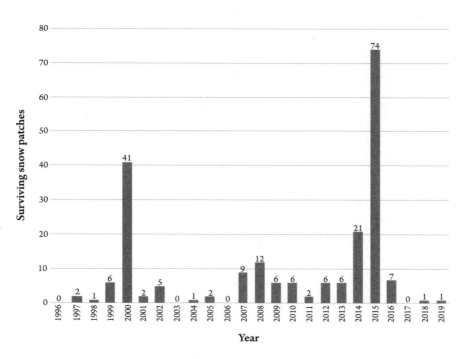

Figure 3: Snow patch survivals in Scotland between 1996 and 2019.
Data from annual Weather journal of the Royal Meteorological Society.

Figure 3 shows that, along with the four years when no patches survived (1996, 2003, 2006 and 2017), there were another eight years when only one or two did. Or, to put it another way, in half of the twenty-four years since 1996, only two or fewer patches persisted from one year to the next. The period of 2017–2019 saw the lowest combined three-yearly total (two patches) since records began, narrowly eclipsing the previous lows of 2003–2005 and 2004–2006 (three patches in each). Never since these snows have been studied have we seen such low numbers.

Interspersed throughout Figure 1 it will be seen that occasionally there is a large spike. Most obvious of all is 2015. In that year seventy-four drifts of old snow made it through the year to the following winter, which is undeniably a remarkable total. Many places that had not seen snow surviving for decades did so again. Perhaps years like 2015 are not such a surprise when we consider that climatologists tell us severe weather may become more common in the future, and that although the general trend may be a downwards one in terms of the amount of snow that falls, big-spike years will endure.

One recent study done on the subject of snow in the Cairngorms predicts that cover will decrease dramatically in the coming years. It states, rather ominously:

> Our modelled future estimates indicate a potential for snow cover in the next decade to continue at a similar quantity to the recent past, with large inter-annual variability. However, from c.2030–2040 there is likely to be a substantial decline in the number of days of snow cover. By c.2050 the trend seen in the past may have continued to the extent that the number of days of snow cover is about half of the long-term observed average. However, variation from year to year, both observed and modelled, suggests the potential that snow cover in some future years may be comparable with past records. The long-term trend is towards greatly reduced snow cover with the possibility of some years of very little to no snow by 2080.[4]

---

4   Rivington, Mike et al. (2019), 'Snow Cover and Climate Change in the Cairngorms National Park: Summary Assessment', www.climatexchange.org.uk [accessed 13 May 2021].

'Very little to no snow by 2080' is a worrying consideration. Scarcely believable, even. How I wonder what those who walked on our hills in the eighteenth and nineteenth centuries would have thought, seeing huge amounts of it even in August, if they'd been told all of it would soon be consigned to history. For my part, the idea of a Scotland without long-lying patches is one which sees the hills diminished. Wreaths of snow in summer give the highest peaks a grace and splendour which elevates them above their modest heights. If their ermine is stripped away by the cruel heat of the sun and the rain, then it will see their beauty lessened. To witness the north face of Ben Nevis from Carn Mòr Dearg in July, with its multitude of gullies filled for hundreds of feet with white, pristine drifts of snow, is to see one of the greatest visual feasts that these islands can offer. How sad it will be for us all to be deprived of such scenes.

Stories from continental Europe, where glaciers are being covered by blankets of foil in order to reduce their melting, have surfaced in the last year or two. I have also heard many people in the UK suggest such a thing for some of our own long-lying patches. My view on this is unequivocal: there's no point. It smacks of desperation. Human nature being what it is, the tendency is to want to *do* something: 'The glaciers are shrinking, we must protect them.' And many have sympathised and been inspired enough to do it. But it is a howl in the wind. All the blankets in the world can only delay the inevitable. This sort of action is reminiscent of the parable of King Canute, where he exhorts the tide to go back out again. It is a futile endeavour. I also dislike the idea of covering a Scottish snow patch up just to artificially preserve it from one year to the next. It feels a little like cheating to me. I have heard people riposte this position with comments like, 'Well, it would only be to counteract the damage mankind has done in the first place.' Well, possibly. That veers too close to the anthropogenic climate-change argument that I try to avoid whenever I can.

If the apparent trend continues on the trajectory it has been on over the last twenty years, it is very difficult to see any other outcome than where patches of snow become the exception rather than the rule. Whether or not that comes to pass remains to be seen, but this scenario is one which no right-thinking person can or should look forward to.

* * *

'Winter'

On the highest Scottish hills, the dates of meteorological winter extremely difficult to define. In theoretical terms, of course, they fall the same dates as a lowland winter (December, January and Februar The reality rarely agrees with the theory, though. In practice, prop winter weather can be encountered on Ben Nevis or the highest of tl Cairngorms in any one of ten months of the year.

Now, it is often said that snow can fall on *any* day of the year or Scotland's hills. This is, however, inaccurate. Certainly, very *poor* weathei can be experienced on any given day of the year. But that is different from *winter* weather. According to unpublished records, since the 1940s there have been nineteen days in August and twelve days in July when it has never been known to snow on the Scottish hills.[5] August, therefore, is the least snowy month of the year in Scotland, followed by July. In all other months snow can be expected to fall reasonably regularly on the highest peaks. But *not*, as is often claimed, on any day of the year.

If snow does fall during the meteorological summer then it tends, though not always, to be ephemeral. Here today, gone tomorrow falls are still relatively commonplace. All it takes is an incursion of Arctic air and some convection across the North Sea to promote showers. After all, Scotland isn't so very far away from the Arctic when one looks on a map. On the rare occasion when fresh snowdrifts *do* persist for more than a few days, then it is usually on only the very highest hills. A good, relatively recent example is 2011. On 28 August that year, a heavy fall, driven on a strong northwesterly wind, caused severe drifting on the Cairngorms. Wreaths from this storm were still visible in mid-September. Fresh August snow has not been known to endure so long as this since at least 1945.

A truly unique occurrence of an early fall of snow on Scottish hills is worth mentioning at length. It is entirely without precedent in the reckoning of people whose memories of such things extend back to the earlier parts of the twentieth century, and it has not happened since.

---

5   Adam Watson kept meticulous records of snowfall in the Cairngorms.

Given the evidence laid out earlier in this chapter, it is debatable if such a thing is ever likely to happen again in the lifetimes of those reading this book.

The event itself happened in 1976. Those old enough to remember that year often wax lyrical about the long days of unbroken sunshine and high daytime temperatures, especially during August. Those of us who weren't old enough to remember it, or had not yet been born, have read about it. It was a summer of very high temperatures, sustained for weeks at a time. Droughts were experienced all over the UK, with water being rationed as reservoirs dried up. Near Aviemore, temperatures of above 20° Celsius were recorded for nineteen days during August.[6] By the end of that month, only two snow patches remained in the whole of Scotland (at the Sphinx and Pinnacles locations in Garbh Choire Mòr, as usual). Both were very small. It was thought that if the hot weather continued then both would not last more than a couple of weeks.

However, the next day, 1 September, fresh snow fell on the high Cairngorms at around 10 p.m. There was, according to one observer, 'a fair covering the next morning at altitudes above 1,080 metres.'[7] Late on the 6th there was more fresh snow, and heavy falls all day on the 7th were accompanied by a north gale. Ten centimetres of wet snow lay at the Lecht and the road summit of the Cairnwell pass beside Glenshee ski centre. During the evening of Wednesday 8th, the weather turned cold again, and fresh snow fell on the high hills. Next day, the 9th, fifty millimetres of rain fell in lower Deeside, which would have equated to snow high up. It has been calculated by I.C. Hudson (as per the previous footnotes) that this would have been equivalent to about sixty centimetres of snowfall. Hudson also reported that fresh snow on 9 September was deposited down to 490 metres on lee slopes due to drifting. In a note sent to me, Adam Watson described the astonishing circumstances and scenes:

6   Hudson, I.C. (1977a), 'Cairngorm snowfield report 1976', *Journal of Meteorology* Volume 2.

7   Hudson, I.C. (1977b), 'Record early cold spell in the Scottish Highlands, 9–16 September 1976', *Journal of Meteorology* Volume 2.

On 10 September, deep snow covered the high Cairngorms but warmer air had arrived on a northeast wind and the snow had started to thaw on the highest tops. My father went to Cairn Gorm that day and reported snow showers and a bitter wind. He walked from the top chairlift round to the hollow of Ciste Mhearad to the east, and wrote in his diary that it was well filled with new drifted snow. His diary also holds his comment that his Norwegian wooden cross-country skis would have been useful. He took some remarkable photographs showing exceptional snow cover. My father told me that he found fresh drifts up to ten feet deep in hollows on the north and east and south sides of Cairn Gorm, and reported deep snow lying all the way down to Loch Avon. A party of Belgian tourists, equipped for summer conditions, narrowly escaped with their lives on 9 September after a two-day visit to Loch Avon, and 'had to be rescued. They were suffering from exposure. If the party had not chanced to meet two experienced mountaineers, at least three would have died.'

Heavy snow showers returned on 12 September, and – all told – it has been surmised that nearly a metre of snow fell during this unprecedented period, with drifts in excess of three metres observed. Adam's note continued:

On 17 September, there were still deep extensive drifts in many places despite warm weather with sunshine. The whole of the southern slope of Cairn Gorm lay under a deep blanket of con-solidated packed drifted snow, which had blown in the gale off the north side. I was on Cairn Gorm plateau on Sunday 19th with my father, a day with bright sunshine and strong glare off the fresh snow. Deep snow drifts lay on the plateau and on the corries of Braeriach, completely covering the old snow in Garbh Choire Mòr. I wrote that the heavy melting and hot sun made it feel like early June. The south side of Cairn Gorm still had a complete cover of deep drifted snow up to three metres in depth, and likewise some hollows on

Cairn Gorm such as Ciste Mhearad. I wished that we had brought our touring skis, for the snow had a firm uniform surface and skiing would have been much easier than walking.[8]

So, what happened to the Sphinx and Pinnacles snow patches that were very small at the end of August? In a trip to Garbh Choire Mòr on 2 October, John Pottie found that much of the early September falls had melted. However, there was still a fair amount visible near where the old patches were located. Upon closer inspection he found that the Sphinx and Pinnacles patches were still lying buried under the remnants of the new falls. The photograph he took shows a birthday cake effect, with the newer snow lying pristine on the top, and the old snow (which by October 1976 was nearly eighteen years old) very much darker and dirtier. Had these new snows not fallen in early September, then it is almost certain the old ones would have disappeared. I have always thought it ironic the best summer of the twentieth century failed to melt all Scottish snow on account of an early blast of winter riding to its rescue.

Whilst September 1976 was an example of a premature taste of winter, another year worthy of note for its unseasonal weather was far more recent: 2015. Though March and April of that year were broadly average in terms of temperature and precipitation, the subsequent three months were anything but. Temperatures in May were 2.5° Celsius below normal, with 175 per cent of the usual precipitation. This translated into very significant amounts of snow falling on the hills in a month that normally sees the snow pack reduce. The cold weather continued into June, which itself had eight days of falling snow over the hills: the most for decades. Pictures of the Ben Nevis range in early June showed the hills more like March, with a thick covering of snow running down their shoulders to well below 2,000 feet. It is quite likely that the deepest snow all year on Ben Nevis was to be found in the first week of June. Spring hardly got started.

---

8   Letter from Adam Watson to author.

The annual August bank holiday survey that I began in 2008 turned into an exercise in damage limitation. For the years before 2015, the total amount of snow patches across the whole of Scotland could be counted in August by about ten people, covering perhaps half a dozen areas. Thirty to forty patches were considered an 'average' total since the survey's inception. But because July 2015 was also cool, a huge amount of snow persisted at the start of August. Anticipating a massive total on survey day, I started desperately trying to recruit volunteers to go to places that I never expected I would ever need to worry about. All told, about thirty people contributed that year, counting a colossal 678 patches across the country, of which 74 made it through to winter. This was the highest total since 1994, and one unlikely to be bettered any time soon. During that year I did over fifty trips to the hills, going through several pairs of boots in the process. As much as I'd like to see a repeat of 2015, I can happily live without another like it for a few years yet.

In conclusion, therefore, it is entirely fair to say that winter is never too far away from Scotland's highest hills, even during these warming years.

* * *

A eulogy to winter

On Friday 10 February several years ago, I received a text message from Alistair, a mountaineer friend.

'A bit last minute, Iain, but with the great forecast and a full moon, I'm heading up to the far north-west tomorrow afternoon, if you fancy a wee snow plod and an amazing-views adventure?'

I checked in my head if I had anything on that weekend. Fortuitously, and unusually, I didn't.

'I'm in,' I replied. 'I'll bring my sleeping bag and will see you at yours.'

Now, the 'north-west' is a huge area, so we needed to nail down the venue. Where exactly would be best to see this snowy landscape lit up, hopefully, by a full moon? Skye was discussed. The view to the Cuillin from Sgùrr na Strì was likely to be unsurpassable. But given its increasing popularity as a viewpoint and the superb cloudless forecast, we thought it might be a bit crowded up there. Places to pitch a tent are scarce, so if we got up there after someone else then it could be a bit of an uncomfortable sleep. The decision made itself in the end: Torridon. And there was only one hill that was going to tick all the boxes.

Liathach is a supermodel. It is a hill of almost unearthly beauty and dimensions. It rises to over 3,000 feet in little under a mile from the road that runs at its foot in Glen Torridon. An air of impregnability surrounds it when viewed from below. A castellated ridge runs along Liathach for much of its length, giving some superb walking and scrambling if one feels so inclined. In winter the ridge is a highly challenging mountaineering undertaking. The seriousness of it increases under snow, but so too does its beauty. Snow embellishes height with most hills, but with Liathach the sensation is magnified even more dramatically. Almost to the point, even, that it can be mistaken for an Alpine summit. Photographs of the hill adorn virtually every calendar that features the Scottish outdoors, and its name pulls on the memory of folk who've been on its many and varied pinnacles. Some hills pass from recollection quickly, but Liathach lingers long. With the Aonach Eagach and An Teallach, Liathach makes up the most revered triumvirate of ridges on the British mainland.[9]

When we eventually pulled up in the car on the afternoon of 11 February it looked glorious. Though meteorologically we were into the heart of winter, only the top half of the hill carried snow. The weak and insipid February sun had little in the way of apricity to trouble Liathach's white veil. It was an inviting view, and soon after 1 p.m. Alistair and I were packed and on our way.

We each carried a sleeping bag and plenty of warm clothing in our rucksacks. The overnight temperature on the summit was forecast to be well below freezing, accompanied by strong gusts. The wind chill was going to be correspondingly fierce. The cold would be exacerbated by having to camp on snow as opposed to solid ground. On this trip, body heat was the currency that was needed to pay for the visual returns.

Despite Alistair's mobility being hampered as a result of a nasty cycling

---

9   Respect is due to Snowdon's Crib Goch and Helvellyn's Striding Edge, but for me they fall just short when compared to this trio.

accident some months previously, we made decent progress up the steep south flank of Coire Liath Mhòr. A well-worn path afforded solid and quick access upwards, even if the going was a little tiring. We bumped into a few people who were descending. They looked somewhat puzzled to see us going upwards at this time of the day, especially with a tent strapped to my rucksack.

It wasn't too long before we reached the crest of the ridge itself. By now we were into thick snow and looking for a suitable camping spot. The sun was still quite well up in the sky and the views were grand. If we could get camped and set for the evening then we would have plenty of time to witness what we hoped would be a memorable sunset.

In order to pitch our tent, we had to clear an area of snow from the only flat(ish) piece of ground visible for what seemed like miles. Specialist pegs were needed to lash the tent into the thick and loose snow. The wind was bitter. It sought out every chink in our clothes. The forecast had been spot on. At a guess it was about -17° Celsius and that was with the sun still in the sky. Eventually, though, we were set. The tent was pitched, and the sleeping bags unfurled; the camping stove was lit, and hot drinks were taken. All we had to do now was to wait about thirty minutes for the sun to set. The moon was already climbing above the horizon near to Beinn Alligin. Everything was set up perfectly. What happened over the next hour I shall never forget as long as my memory holds.

As the sun dipped behind the westernmost peak of the ridge, Mullach an Rathain, the sky began to reflect its dying light. A palette of reds and yellows and blues and purples coated the clouds. The nearby peak of Beinn Eighe was set ablaze by the low-trajectory shafts of sunlight. One of its summits, Sail Mhòr, was cut in half by the shadow of Liathach. On the bottom part, semi-darkness; on the top, an iridescent pyramid of sandstone. The snow that lay on the summit cones of the surrounding hills sparkled with a golden light as though layered with gilt. Despite having seen snow in many guises over the years, I had seldom known it to turn such a colour. Further afield, in this ragged edge of the country, the shattered, ancient sandstone edifices of Torridon were coated in a blood-red hue. The magnificence of the colour was exacerbated by the rich white blanket of pristine snow that sat atop it. In over one billion years, the age of this landscape, surely it had never looked so good.

Three hours before, the weak February sun that shone overhead had struggled to add any definition to the prehistoric peaks. Its azimuth was too high. Now, as it sank beyond our vision, it concentrated all its magic on the thousand-million-year-old hill, serving up a visual treat the like of which can surely only be witnessed on a handful of days every few years.

Finally, the sun did set. The purple and blue hues that illuminated the clouds faded. On the horizon, though, colours changed from deep blue to a glowing orange as the light refracted off the stratosphere. Crystal-clear air revealed the Cuillin of Skye in the distance. So sharp was it that individual peaks could be discerned even with the naked eye. Eventually the orange faded as well and was replaced by dark blue, then black. The light show put on by the sun had ended. We sat on the snow, dressed head to foot in every fleece and coat we had brought up with us. We watched the very last of the colours empty from the horizon. It was now punishingly cold with no sun left to draw any warmth from. Phoebus had now gone for his long winter slumber.

We stood up and walked around a little to ward off the bitter gusts that by now brought with them wind-chill temperatures in the minus twenties. Although the sun had gone and the sky was dark, we had superb visibility. This was due to the full moon overhead lighting up the snow as though it were floodlit. In the distance, Beinn Alligin's snow-capped horns shone almost as brightly as they would at sunrise. Liathach's burnished peaks stretched skywards, the snow picking out each rock and cliff. Moon-shadows abounded. I had in the past walked on snow that had been illuminated by a full moon, but never in air as clear as this, nor with such views. The experience was almost transcendental. We took pictures to try and capture the moment, but they were a pale imitation of the real thing.

In due course we were beaten by the cold and had to call it a day. We scurried into the tent and settled down for the long night ahead. Never have I been so glad of a sleeping bag. During the night, as I lay waiting for sleep to arrive, I thought of what our experience would have been like had our trip been done without snow on the hills. The clouds may have been just as pretty, but the hills would have been diminished. It may have been physically more pleasant to be there in summer, but for the memories it bestowed on us winter was unbeatable.

* * *

Climate change may yet claim Scotland's winter. Many people will say that the loss of snow from our hills isn't nearly as pressing an issue as the havoc that may be wrought across the world by its effects. That may be true enough, and I do not wish to make light of the troubles that might accompany these potential changes. But a loss of snow on our hills is not without consequences either. A privation of biodiversity and civic amenity can be anticipated, and that's just for starters. Wildlife is sure to suffer just as businesses are certain to fail.

Many walkers live for the day when they open their curtains and see the ground and trees coated in thick drifts. Snow not only transforms the landscape, it transforms us. It has the power of atavism, to turn normally sensible adults into playful children, drawing out the dormant magic that many have lying deep within their souls. One only needs to look outside on a cold, wintry Sunday morning to see and hear the smiles and laughter of parents sledging down hills with their sons and daughters. Online, many folk become unaccountably excited even at the prospect of snow *falling*. No other regular weather event causes such fervour and derangement.

One can only hope that the predictions of climate-change doom do not come to pass, for all our sakes.

eight
The grand tour

We are fortunate in this relatively small country of ours to possess so many different and distinctive regions. From the Ochil Hills near Stirling through the Trossachs and down to the Southern Uplands, the snow-spotter's lot is a varied and happy one. In this chapter, I list some of the finest places in Scotland to visit old relics of winter snow. By adding a map reference of the snow positions, I am not necessarily seeking to encourage others to go there. Rather, this allows the reader to see the spot on a map should they so desire. Any literature that encourages someone either to explore for themselves or to open a map can only be a good thing.

* * *

Aonach Mòr (snow location: NN193736)

Aonach Mòr sits at Scotland's top table. It is a hill bettered in height only seven times across the country, topping out at 4,006 feet. But it is not just for its height that it deserves to be in such exalted company. Uniquely amongst West Coast hills, one can walk for two miles in a straight line and not once dip below an altitude of 3,000 feet. This may seem like a trifle, but to those of us who enjoy a bit of respite from the West's constant up and down ridges, it is bordering on blissful. Not even Ben Nevis can boast such constant terrain. This is but an aside, though. Its credentials for being at the top table are many and varied.

Speaking of terrain, in this regard it is a peak of contradictions. Whereas the summit of virtually every Scottish hill in excess of 3,500 feet is a jumble of stony chaos, Aonach Mòr's is a study in green serenity. Thick jade and olive-coloured carpets of grass and rare mosses spread across the plateau to the extent that the walker would be forgiven for thinking they were striding across Devon pasture, rather than on some of the highest land in the UK. Camping on this verdure is a little treat that I give myself every now and again, especially if it's on a still summer's evening, high above where even the most intrepid midges venture. The thick moss that pockets some parts of the summit plateau is so soft that no camping mat is necessary to facilitate a good night's sleep.

Another contradiction is its name. Aonach Mòr in Gaelic means 'the big ridge' or 'the big ridged hill'. However, it is not as high as its immediate neighbour, Aonach Beag, which means – conversely – 'the small ridged hill'. The Gaels who named it were patently referring to the shape of the hill rather than absolute height. (Either that or they had a rather wry sense of humour.) Its shape is pleasingly uniform from the glen below when viewed from the east. The summit ridge is, as mentioned, relatively flat for quite a distance.

Guarding the plateau on both the east and west sides are a series of insurmountable cliffs and gullies. The only realistic access the walker enjoys is from either the north or the south. Come from the former, though, and one is obliged to negotiate the ski centre paraphernalia that reaches to the very edge of the plateau. Come from the latter and one is compelled to ascend via a rather precarious and indistinct ridge which, though doable without too much technical difficulty, provides a stern test for those who suffer with a fear of heights (this includes the author).

As part of writing this chapter I tried to count how often I have been to the summit of this hill. I gave up quickly but reckon it must be close to fifty. The reason for this multitude of visits is that Aonach Mòr is home to one of Scotland's longest-lasting snow-patch sites. Moreover, in its great north-east corrie, Coire an Lochain, lies one of the country's most interesting and unusual locations.

If one were asked to draw a picture of the stereotypical Highland corrie, Coire an Lochain ('the corrie of the small loch') would be the outcome. Tall cliffs fall away from the summit plateau for a couple of hundred feet before giving way to steep, stony ground that angles towards the floor of the glacial trough. At its base lies the eponymous small loch. All told, from summit to lochan, 1,200 feet separate top from bottom.

Because of its great height, in Scottish terms, Aonach Mòr sees a lot of snowfall. Being the largest of the corries, Coire an Lochain gets a fair proportion of that snow. Consequently, the bravest skiers and snowboarders flock here in spring to put their skills to the test against the challenging terrain of Coire Dubh, Easy Gully (a misnomer if ever there were one) and Summit Gully. Hidden from view for much of winter, spring and even sometimes in early summer, there lies in Coire an Lochain Scotland's most impressive (in my opinion) periglacial feature.[1] Namely, a protalus rampart. This grandiose title can be explained simply as:

a simple landform. It requires the existence of long-lived snow banks below rock cliffs. Frost weathering leads to rockfall from the cliff and the blocks roll down the snow slope to accumulate at its base.[2]

Like so many of the long-lying snow locations that are described in this book, it is a demanding place to get to and hardly ever visited in summer. As a result, it retains a real sense of isolation and wilderness, despite its reasonably close proximity to civilisation. In large part this isolated feeling is to do with what the walker finds when he or she gets there. It is a beguiling spot. When one visits during a 'normal' summer in, say, July then it is possible to peer into a glacial past. Stand by the rampart, similar to that shown in the diagram overleaf, and one is standing amidst an evolving landscape, shaped by the elements to this day.

1   Of or relating to an area or feature close to an ice sheet or long-lying snow, which influences the climate or geological processes of that locality.
2   Further details about protalus ramparts may be found at https://gh.copernicus.org/articles/70/135/2015/gh-70-135-2015.pdf [accessed 13 May 2021].

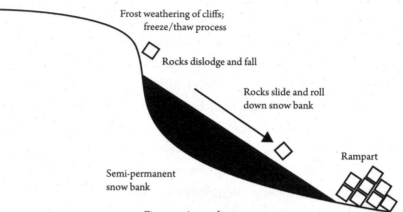

Frost weathering of cliffs;
freeze/thaw process

Rocks dislodge and fall

Rocks slide and roll
down snow bank

Rampart

Semi-permanent
snow bank

Figure 4: A protalus rampart

On the cliffs above the long-lying snow patch, water percolates into the fissures and cracks of the granite, either because of winter rain or snow-melt. Once this water freezes it expands in the joints and renders the granite unstable. Eventually, by the slow hand of erosive time, a piece of this cliff will surrender its grip and fall. Waiting below, the large patch of long-lying snow acts as a chute, deflecting the flying boulder to the bottom. Over many years the pile of stones gets larger and larger, becoming a rampart many metres high. In fact, it perpetuates its own success. The larger the rampart gets, the deeper the hollow that it protects becomes. More snow is added the following year and the cycle is repeated. On and on it goes. Each year, slowly but very definitely surely, the rampart's defences are added to by the presence of the snow it inhibits.

This landscape is not static. It is in a constant state of change. Imperceptible differences accumulate over hundreds and thousands of years. Our own fleeting time on the planet allows us but a brief glimpse of this relentless process which will, ultimately, turn the hill back into the dust it was fashioned with.

* * *

Geal-Chàrn (snow location: NN480748)

Though it sits far closer to Ben Nevis than Ben Macdui, Geal-Chàrn's appearance – like that of its immediate neighbour Ben Alder – is definitely more akin to the Cairngorms than to Lochaber. Both of these magnificent hills have high plateaux which rise gradually from the west, only to drop vertiginously into east- and north-east-facing cliffs. Of the two hills, Ben Alder takes most of the awards. In height it has an extra sixty feet. In visual splendour it surpasses virtually every hill in Scotland, let alone Geal-Chàrn. It also contains one of Scotland's highest named bodies of water: Lochan a' Garbh Choire – at just shy of 3,700 feet. Geal-Chàrn, however, has Aisre Cham.

Aisre Cham – pronounced, improbably, as 'ashra CHOWM' – means in Badenoch Gaelic 'crooked pass' or 'crooked ridge'. Why it has this name is a mystery. No ridge or pass exists here, so its true meaning will probably stay lost in the haze of time. The Aisre Cham is, in effect, a hollow that sits just below the eastern edge of the massive plateau of Geal-Chàrn. Like so many other long-lying snow locations it benefits from windblown snow piling into it over winter and spring. It is my view, though I stress this is just a hunch and based on nothing in the way of science, that this site is one that can rival even Ben Nevis for sheer depth. In the snowiest years I have seen accumulations there that bury virtually every rock and crag. When one ventures there in late summer or early autumn one realises just how much snow it takes to do that. We are talking in the scores of feet deep here. And what an expedition it is to get there!

The 'Ben Alder round' in Scottish mountaineering circles is considered a classic. Before the closure of Culra bothy on health and safety grounds (it has an asbestos roof), most weekend warriors would walk there on a Friday evening from Dalwhinnie before attempting the six Munros the next day. Because the walk back was ten miles, they might even decide to stay another night and saunter home at leisure on the Sunday morning. With that option no longer available, walkers must either take their own tents, sleeping bags and camping equipment in; or, as is becoming increasingly common nowadays, do the whole thing in a day, travelling from and back to Dalwhinnie by bike. I first did this round in 2012, and I used a bike. It is a huge day, not recommended unless one is very fit. The three other times I have journeyed back and forward to this range

I have also used this method of transport, but eschewed doing the six Munros. Whenever I venture there nowadays it is to the Aisre Cham I go. The snow patch there is one of the few that has been known to survive from year to year outside the Cairngorms or Ben Nevis range. For that reason alone, it is worth going. I also find the location grand and wild in a way that only the most remote of hills can be.

During a memorable visit of 12 October 2014, I went to the Aisre Cham via a ridge called the 'Lancet Edge', which connects Geal-Chàrn with its lower neighbour Sgòr Iutharn. The clue to the memorability of the visit is tied up in the ridge's name. One section of the razor-like ridge almost got the better of me. Not for its technical complexity – it was a simple step – but, rather, its exposure. Not even on the famous Aonach Eagach ridge did I feel so much trepidation as I did there. My old nemesis of being afraid of heights made my head swirl at the prospect of stepping over this minor gap in the ridge. Perhaps the fog of time has made it seem more difficult, but my memory of it lingers so vividly that I will not attempt it again. At the time I remember making a pact with myself that this was the first and last time I would attempt it. The reward of seeing two ptarmigan on the rocks just as I'd crossed the gap went some way to making up for it.

As to the location of the Aisre Cham itself, it is the classic long-lying snow hollow. Vegetation is in short supply, with the usual collection of primitive liverworts and mosses huddled in little clumps wherever possible. The ground is perpetually wet due to snowmelt, even during the very driest summers. It is tricky to access and, when one does, the moving, waterlogged ground means one has to tread with caution.

But it is worth the effort. How well I remember walking off the summit of Càrn Dearg on 1 October some years ago, in wonderful autumn sunshine. As I descended on soft, dry grass I could see Geal-Chàrn ahead of me. Two large patches of snow glowed in the early afternoon sunshine at Aisre Cham. I plodded at a leisurely pace toward them. As the ground steepened along the twisting ridge of Aisre Ghobhainn it felt like I was hardly expending any energy, my feet moving almost of their own accord. The heat of the sun was on my back and time didn't count for anything.

Reaching the snows, I sat beside them and looked out across the barren expanse of the Ben Alder deer forest. It was little short of magnificent.

* * *

Glen Almond (snow locations: NN771370 and NN786371)

In an isolated part of southern Perthshire sits one of Southern Scotland's finest glens. That said, if popularity were a measure of how fine something is, you wouldn't know it if you were to walk along the Land Rover track that runs through it. Even on a weekend you might only come across a few mountain bikers or the occasional walker. In point of fact, I have encountered more people fishing here than any other activity. I know why. Munros. Or, rather, the lack of them.

Glen Almond benefits enormously from having no hills higher than 3,000 feet. Technically Ben Chonzie qualifies as being accessible from the glen, but in reality no Munro-bagger is going to come at it from this direction, for to do so would mean a massive diversion. Munro-baggers generally take the path of least resistance. This lack of high hills means that one can walk there in relative peace and tranquillity, save for the odd estate vehicle (more on this later, alas). It is also a cul-de-sac. Many glens act as conduits through the Highlands and as a result have attracted through-roads. Not Glen Almond. In this respect it is like Glen Lyon – a princess amongst Scottish glens. There are few houses to be seen, and those that are visible often seem to be unoccupied. Auchnafree House, the large estate building, seems to be the population hub of the glen.

When I moved back to Scotland in 2014, after having lived in England for the best part of ten years, Glen Almond was the first place I decided to camp in. Its wildness, relative proximity to my house in Stirling and lack of footfall were the main reasons for this. The other, as you may now have guessed, was to see some snow.

Apart from the peacefulness of the glen, which is certainly one of its main draws, the unusually flat terrain aids the walker or cyclist. In nine miles, from the glen's eastern terminus at Newton Bridge, the ground rises only 300 feet. In so doing the road passes a host of interesting historical features like the Neolithic chambered cairn Clach na Tiompan and its nearby standing stones, the Stuck Chapel and millennium memorial, and – of course – the wonderful crags and cliffs above on either side of the glen.[3]

A host of estate roads peel off from the valley floor to attain the summits of the hills on the northern flank of the glen. These are predominantly to aid the 4x4 vehicles taking shooting parties up to the innumerable grouse butts that exist all over the hills here. The best, and the steepest, of these roads is the one that runs up the shoulder of Sròn an Fhèidh ('the point of the deer') on to Sròn a' Chaonineidh ('the point of mourning'). Any pretensions of staying on one's bike for most of the traverse upwards are laughably curtailed within a minute or two. Even an electric bike, I fancy, would toil for grip on the relentlessly steep 'nose of the deer'.

Sròn a' Chaonineidh is the central hill in a line of three that consists of Creagan na Beinne to the west and Meall nan Eanchainn to the east.[4] All three have pronounced northern corries that, despite their relatively modest height, can gather huge volumes of snow and which often persist far longer than hills of greater height. Sròn a' Chaonineidh's Coire Quaich is the finest of the triumvirate. The best time to visit this corrie is in mid- to late April. If the winter has been snowy, extend that by a couple of weeks. By a distance the most impressive I've ever seen these snows was in June 2015, when I took a tent up and camped on the lush grass that carpets the wide summit ridge. On that visit I was expecting to find some snow, but not a huge amount. How wrong I was. Upon getting there the whole rim of the corrie, for hundreds of metres, had its own glacier, complete with mini crevasses. I climbed down into a couple of them and walked along for quite a distance. This made up for the camping disaster I had that night by forgetting to pack the tent pegs. (To cap it all it started raining. Were it not for military tarpaulin I threw in at the last minute it would have been a thoroughly miserable evening.)

---

3   Technically the glen carries on through the Sma' Glen and past Glenalmond College, but for the purposes of this chapter it is the section west of the A822 we are concerned with.

4   Creagan na Beinne is another name the Ordnance Survey got wrong. It's pronounced locally as 'crehgan a VEHYn'. Meall nan Eanchainn means 'the hill of the brains', according to Dwelly's *Gaelic–English Dictionary*, Glasgow: Akerbeltz Publishing.

Slightly farther east, Meall nan Enchainn's equivalent corrie is Coire nan Gabhar.[5] For some reason, probably that the walls of the corrie are a bit less steep, the build-up of snow here is usually a bit less, but it is still a fine location.

Without wishing to labour the point too much, these hills in spring are – for their isolation, peacefulness and snow – among the very best that the Southern Highlands have to offer. They are also a superb place to see certain wildlife. A lack of walkers in general means some birds abound. Golden plover in particular are common, their shrill cries a pleasure to hear in the cool breeze of the summit ridge. Mountain hares – *Lepus timidus* – are often seen, but more skittish. If one is lucky, eagles may be viewed soaring overhead.

All of that said, recent events on these hills have soured their appeal for me. In 2019, a golden eagle went missing on the estate that these hills lie in. Its satellite tag stopped functioning after it spent an inordinate amount of time on the ground. To compound this the eagle itself was called Adam, so named by member of the Scottish parliament Andy Wightman after he adopted it. The Adam in question was – sadly – Adam Watson. Andy, like so many others, looked up to Adam and had a huge respect for his work. (I know Adam was delighted when Andy named the eagle after him.) The exact location of the disappearance was narrowed down to the southern spur of Sròn a' Chaonineidh, right at the end of an estate road. Like so many eagle disappearances on managed grouse moor, the whole thing is remarkably suspicious. I avoided the area for a while afterwards because of this. It left too bitter a taste in the mouth.

Of course, it's not the hills' fault that this happened. But, humans being the irrational creatures they are, sometimes take things like this a bit too personally. I will be back soon enough to see those hills, and their wonderful snowdrifts.

* * *

---

5   Coire nan Gabhar, 'the corrie of the goats', is pronounced as 'carn nan GO-er'.

Ben Macdui (snow location: NH994014)

The Feith Buidhe ('the yellow bog stream') is a small and short burn that has a big story to tell.[6] Its water is not yellow, as the name might suggest. Rather, it is the pale yellow mat-grass that is to be found adjacent to the stream for much of the year that gives it its name. The source of the burn is the Lochan Buidhe ('the small yellow loch') which, at 3,700 feet, happens to be Britain's highest body of named water.[7] The Feith Buidhe runs for only 1.5 miles and empties into Loch Avon.[8] The journey cannot take the water more than an hour to complete. Seldom has such a short route encountered so much visual beauty or tragedy.

Regarding the latter, much has already been written about the dreadful Cairngorm plateau disaster of November 1971. Its alternative name is the Feith Buidhe disaster. For those not familiar with the awful story, it involved a group of six fifteen-year-old schoolchildren and two leaders. Of the eight people in the party, five of the six children and one of the two leaders were killed when they decided to open-camp near the Feith Buidhe in horrendous winter weather. They had been heading for a hut, but – due to the atrocious conditions and lack of experience – never made it and most perished after the disastrous decision to bivouac in a blizzard. The group was but a few hundred metres from the shelter they were aiming for, which makes the story even more tragic. In summer I often pass the site of their bivouac. In benign conditions it is scarcely believable that such an attractive spot can turn so lethal.

A few hundred metres downstream from this solemn site, the Feith Buidhe tumbles off the plateau and down what locals used to call Creagan na Feith Buidhe ('rock of the yellow bog stream'). Once it has crashed down these granite terraces it joins with several other streams and morphs into the wonderfully named Meur na Banaraich ('the finger stream of the milkmaid'). This short, new river has barely had enough time to settle when it is subsumed by the queen of all Cairngorms lochs: Loch Avon.

---

6   Feith Buidhe is pronounced as 'fay BOOee', with a thin 'B', almost like a 'P'.

7   Lochan a' Garbh Choire on Ben Alder is the other claimant to this title. Both sit at 3,700 feet, but Lochan a' Garbh Choire is more a puddle than a loch and has no outflow.

8   Pronounced as 'loch AAhn'.

The granite terraces the Feith Buidhe tumbles down are no longer called Creagan na Feith Buidhe. They are now styled (the) 'Feith Buidhe slabs'. The old name has all but passed from use. The view down Loch Avon from the top of these slabs is one of the classic visions of the Cairngorms. Personally, I consider it to be bettered only by the one from the summit of Sgòr Gaoith across to Braeriach. The Loch Avon view is enhanced considerably when one ventures there in early or mid-summer. Huge volumes of snow gather on these east-facing slabs in winter and spring. The plateau to the west has no major dips or hollows to contain the snow's eastward journey. When it blows over the edge it gathers on the neatly symmetrical slabs and piles high. Very high. There can be fewer places in the Cairngorms where so much gathers, in fact.

Like many fascinating snow-patch locations in the Cairngorms, the site is missed by almost every walker. Maybe this is down to the snow being invisible from the path that connects the two highest points on this upland shelf: Ben Macdui and Cairn Gorm itself. I doubt that's it, though. Even if it were not invisible, it is unlikely that many would bother to make the detour. Most are intent on 'bagging' the Munros in an already long day's walk. It is their loss, for at the slabs – in the right summer conditions – are some of the greatest snow sculptures that Scotland has to offer. Certainly, they are among the most visually impressive. In this regard the Feith Buidhe slabs themselves are the architects.

Imagine a series of terraces. Think of the stepped pyramid of Djoser in Egypt, or rice paddies on a hill in Vietnam. Now imagine them covered entirely in snow during winter. When spring arrives, the snow starts to melt from the top. A few warm days in May expose the top terrace, and the uncovered ground absorbs the heat produced by the sun. As this ground warms it starts to emit heat in exactly the same way a household radiator does. What happens next is that a gap appears between the top of the terrace and the snow. This gap is now the weak link in the chain. Warmer summer air starts to pour in and eventually reaches the bottom, opening a diagonal gap between the terraces and the snow. To help visualise this, imagine cutting a block of cheese or butter in two. But instead of cutting it straight down the middle, start from the top right edge and slice it diagonally to the bottom edge on the left. Two wedges have been produced. Slide them apart an inch and you have the scene at Feith Buidhe slabs in summer. The left wedge represents the snow, and the right wedge the granite terraces. But this is where it gets *really* interesting.

The space between the snow and the granite terraces is such a rare and little-seen phenomenon in the UK that there are no English words for it. Instead we need to look to the continent for suitable nouns. The Germans call this space '*randkluft*'.[9] In June and July this gap is often wide enough to admit the passage of several people at a time. *Randkluft* is rare even in Scotland, but it is nowhere better seen than at the Feith Buidhe slabs. The results can be extraordinary.

The terraces of the Feith Buidhe slabs are tens of metres high. Each shelf is made up of grippy, solid granite. When one walks on to the topmost terrace, then clambers down the rest of them to where the foot of the snow is, it is like nothing else in the country. The first thing one notices is the huge wall of blue ice that one is standing beside. So high is it that it's impossible to gauge its depth. It slopes over one's head and away, angling and stretching away from the walker. The ice's foot is rock hard, as though glacial. It has been compressed by tens of thousands of tonnes weighing down upon it and pressing against the ground. The air is cold there, many degrees below the outside ambient temperature. So much snow and ice are there that the very atmosphere has changed. This is the deep heart of the *randkluft*; the place with no English name. The blue and green hues of the snow and ice's fringe project a science-fiction feel. Shouting at the top of one's voice, a weird and unsettling echo rebounds off the terraces and snow. It is altogether a disconcerting place to be, albeit one of stunning and ethereal beauty.

But it is also dangerous. Though in the deepest bowels of the *randkluft* the risks are minor, at the fringes, where melting is most pronounced, danger lurks. Several years ago, as I traversed the slabs, looking down at a suitable place through which to enter the yawning *randkluft*, I saw some precarious and unsupported snow at the edge of the huge drift.

---

9   Meaning 'marginal cleft or crevasse'.

I declined to go down to it for fear of it collapsing, instead opting for another mode of entry. No sooner was the thought out of my head than the unsupported snow gave way, taking a huge chunk of the whole pack with it. From a distance of twenty metres or so I felt the air pass across my face, as though a great fan had been wafted. Perhaps a hundred tonnes of collapsed snow lay on the slabs, and had I been in the vicinity I would no longer be alive to recount the story. As can be the case in life, beauty is often accompanied by danger.

* * *

Beinn Ime (snow location: NN256084)

This peaceful peak ('the butter hill') is an altogether more sedate affair than most other hills that I visit regularly to ascertain the condition of snow. The journey to get there is short, lying around an hour's drive from Stirling or Glasgow. If one is smart, one will avoid the ascent up the hill from Arrochar and its car parks, which attract hordes of people wishing to walk up The Cobbler – perhaps Scotland's finest hill under 3,000 feet. There is also the matter of the not inconsiderable parking charges that Argyll and Bute Council have now started to levy on the motorist at The Cobbler car park. Being a canny Scot, such taxes on hillwalking are always avoided if I can possibly get away with it.

Mostly, for me, however, it is the hordes that I try to avoid rather than the cost. And for that, one should start on the north-west flank of the hill, close to the junction of Glen Kinglass and the A83 'Rest and Be Thankful' road. From there it is only a 1.5-mile walk to the summit of Beinn Ime – pronounced as 'ben EEm'. This distance, too, is more sedate than some of the mega-walks required to get to other long-lying patches. I say 'sedate', but this – like 'long-lasting' – is a relative term. One must ascend over 2,500 feet in those 1.5 miles from bottom to top. That can quite reasonably be categorised as 'lung-bursting' in most people's reckoning. It is unforgiving, though brief. Do not tarry in summer, mind. It is a place of biblical swarms of flies and – worst of all – clegs: the beasts of the hill. Midges are but an annoyance compared to the bloodthirsty cleg. Even the way that Scots say the word (with a very thick *cl*) is onomatopoeic. It is spat out.

The summit of Beinn Ime is the highest point in the so-called Arrochar Alps, standing at a proud 3,316 feet. Arran can readily be seen to the south-west, as can Ben Nevis to the north. The Paps of Jura light up the western view, and on a *very* clear day with binoculars it is apparently possible to see Sawel Mountain, the highest of inland Northern Ireland's hills – at a distance of 131 miles. I have yet to have such clarity. (On that note, I have spent a little time researching the theoretical longest lines of sight in the UK. The longest I have seen postulated is between Carnedd Llewelyn in Wales and Cairnsmore of Carsphairn in southern Scotland, a distance of a scarcely believable 145.3 miles.)

My annual jaunt to Beinn Ime usually happens in late May or early June. By that time, it generally carries the UK's most southerly patch of snow. That particular baton is handed to it by Ben Lomond, which – in turn – is passed it by one of the Moffat hills in the Southern Uplands. That being the case Beinn Ime is one of the bellwether patches and a close eye is kept on it.

On a blisteringly hot day in late May 2017 I went up Beinn Ime with a Stirling University professor and one of his students. They were conducting some research on vegetation and its relationship to long-lying snow. (At this particular patch site, the vegetation spends anything up to six months of the year buried by snow.) On the way up the heat was oppressive. The car thermometer showed 28° Celsius, and it felt every bit of that. The fair-haired student toiled on the upward walk, unsurprisingly, and even when we reached the summit it was still over 20° Celsius, according to the calibrated thermometer we had with us. The snow that we had come to check was pitifully small, far more so than in previous years. But it was snow nonetheless. The professor and student carried out their experiment and took some data. The heat continued for the downwards walk. When we eventually started the drive home the student suddenly projectile-vomited profusely all over the inside of the professor's car. The professor showed a remarkable calmness despite this. The heat had got to his student, badly. It is the only time I have seen heatstroke when conducting a snow survey. Usually the worry is at the other end of the temperature spectrum.

The location of the snow itself is in a classic north-east-facing basin below the summit cone. It gathers readily and fills to a great extent. In 2015, I went there in July and was astonished to find a good deal of depth still present. It is the only year I have known where it lasted into August: unheard of for such a relatively low hill so far south. For me, the visit to Beinn Ime in late spring signifies the start of the snow-patch-spotting season *proper*. I look forward to my next trip with relish.

* * *

Ben More, Crianlarich (snow location: NN432246)

There are steep hills, steeper hills, and then there is Ben More. Its summit – like Beinn Ime's – is easily accessible from the roadside, a mere 1.5 miles away. Only, pile another 500 feet on top of Beinn Ime, with no added distance. If one goes straight up it from the A85, several deep breaths are advised before setting off. Once one hits its main slope, the angle and effort are unrelenting.

Ben More has the simplest of Gaelic names – 'big hill'. From some angles it looks elegant: conical, smooth, enticing. From others it looks like a bruiser: wide, muscular, solid. With its inseparable partner, Stob Binnein, it dominates the landscape around Crianlarich, the small village where the roads from Glasgow and Edinburgh to the West Highlands meet.

Of these two high peaks (the sixteenth- and eighteenth-tallest in the UK), Ben More is the higher by thirty feet, at 3,852. However, Stob Binnein – 'the anvil peak' – is the shapelier. The former is devoid of the graceful edges and rounded curves that the latter possesses. To see Stob Binnein's crest stretching up to the hill's summit from the top of Ben More is to witness one of the grandest sights in the Southern Highlands, especially when covered in snow. Handsome cornices creep out from the ridge's edge, creating a scalpel-like rim almost to the summit. Discounting Ben Lawers, there are no higher hills farther north until one gets to Ben Nevis. Because of this great height, Ben More – especially of the two – enjoys a lot of snow. All aspects of this hill can receive large amounts because of its conical nature.

In early March 2010, at the end of a very long and cold winter, the aftermath of a huge avalanche was seen on Ben More by a couple of walkers. I recall being informed of it at the time, but because I was living in England I was unable to attend. Fortunately, the walkers who witnessed the debris took many photographs and posted them online for others to view.[10] To my mind it is the largest and longest single avalanche event in Scotland for many years. For the most part the slide was caused by the hill's steepness, allied to the amount of snow that was present on the hills in general that year. The pack would have been poorly bonded (by a lack of freeze/thaw cycles), therefore when the temperatures warmed up slightly the snow was highly unstable. Such was the force of the Ben More avalanche that it 'blasted' its way across the opposite side of the small glen and had started to travel *upwards*. Events like this are rare in Scotland and are reminders that in full winter condition the hills need to be treated with a great deal of respect.

The main place of interest for the chionophile on these two hills is high on Ben More's north-east shoulder: the Cuidhe Chròm.[11] It is unique in Scotland in that it is the only place where a snow patch is named on an Ordnance Survey map. The Cuidhe Chròm ('the bent or curved wreath') is incredibly conspicuous in late spring and early summer for many miles, adopting a pronounced boomerang shape as it crests round a steep slope below the summit cone. It shares its name with two other Cuidhe Chròms in Scotland: one on Lochnagar and one on Cairn Gorm, though the latter is spelled *cuithe* rather than *cuidhe*. The name is evidently hundreds of years old, stretching back to when Gaelic was the spoken language in this area. The appellation was established enough to be included on a map from 1888. Its conspicuousness is almost certainly the reason why it is thus named.

The location itself lies very close to the path that hillwalkers use to attain the summit and it is one of the most accessible long-lying patches in the country because of this. It is, though, an incredibly steep place to visit. A slip here would have grave consequences, almost certainly

---

10   'Crianlarich Avalanche': thepeeler-lawrie.blogspot.com, available at: http://thepeeler-lawrie.blogspot.com/2010/03/crianlarich-avalanche.html?m=1 [accessed 13 May 2021].

11   Pronounced as 'KOOee chrome'.

involving tumbling many hundreds of feet on to rocks below. Another few degrees of steepness would render the location impossible to visit without specialist gear. So steep is it, in fact, that I have used crampons on my boots even when on the grass that surrounds it.

On a glorious, warm May afternoon it is a special place to be. During dry spells the melting snow is a constant source of water. As a result, insects buzz on the fringes of the snow and birds seek it out for food and water. Ptarmigan and eagles are frequent visitors to the vicinity. As well as the wildlife, the views eastward to Loch Tay are sublime, as are those north towards Ben Nevis.

The Cuidhe Chròm is one of the most visible and observed patches in Scotland and as a consequence, excellent melt-date data exists for it. Every summer driver heading into the West Highlands from the south-east of Scotland, on a clear day, sees the Cuidhe Chròm. I wonder how many of them realise its significance.

* * *

Ben Cleuch, Ochil Hills (snow location: NN901005)

Nae silks or satins I'll put on,
Nae flooers shall bloom for me –
But Lady Alva's snawy web
My winding-sheet shall be.[12]

The Ochil Hills are special to me. Not for their height, their cliffs, their wildlife or their drama. They possess none of these attributes in any comparable way to the dramatic peaks of the Highlands. Also, if I'm honest, there is much about the Ochils that I would like to change (there are far, far too many sheep on the hills, for example, and too few trees).

The reason the Ochils mean so much to me is that they are my local hills. I see them every day at close quarters and as such I know their shapes and outline intimately. It is my go-to range when time is short or

---

12 Crawford, John (1850), 'Katie Glen', *Doric Lays: Being Snatches of Song & Ballad*, Edinburgh: Alloa [accessed via www.electricscotland.com/culture/features/scots/doriclaysoocraw.pdf].

long travel is unappealing. I have been to their summit, Ben Cleuch, over a hundred times in all weathers. I love their grassy, steep slopes. Rain drains from the southern flanks of these hills like no others. No matter how hard the torrent, give it twenty-four hours and it's possible to walk up in training shoes without getting wet feet.

Because they sit close to the edge of the Highlands, as well as some of the major population centres in Scotland, they offer almost unrivalled views of both on a clear day. Arran is visible from Ben Cleuch's summit (just 2,365 feet), as is Ben Macdui – just. The panorama from here also extends south to the Pentlands, which are visible above the three Forth bridges. If one looks closely, Berwick Law and the Bass Rock can also be seen.

The normal route up to the summit of Ben Cleuch is fairly steep, but only in places. This upwards path, via Silver Glen and then Ben Ever, prohibits viewing the summit cone until one is about two-thirds of the way into the journey. When it does come into view, assuming the walk has been undertaken in early March, the viewer will not only see the upmost point of the Ochils, but also the charmingly titled 'Lady Alva's Web' snow patch, as referenced in the opening poem.[13] The origin of this name is lost to history. Was it after Lord Erskine's wife, or is Lady Alva just a female name for the town itself? Whatever its history it lends a nice description to this fragile wreath of snow.

It is formed only in the right conditions. If the predominant weather in winter is from the Atlantic then the wreath will not develop, irrespective of the amount of snow that falls. Unlike most well-known snow wreaths, Lady Alva's web faces south-west. Therefore, it forms most fully when Arctic weather systems predominate. In 2010 it was massive, exposing multiple dead sheep when it finally melted out in May of that year. From that year it has formed only sporadically and – even then – is typically ephemeral because of its height and aspect. But if one can go quickly to it after it forms, the wreath is highly impressive for its bulk. The topography of Ben Cleuch means that big accumulations are formed quickly.

---

13 It has also been called Lady Alva's 'diadem' and 'necklace', but 'web' is best known locally.

Gazing down to the remains of Alva House, it is just possible to imagine the eponymous Lady staring back.

* * *

Creag Meagaidh (snow locations: NN428883 and NN428882)

Harold Raeburn was a man apart. Born in Edinburgh in 1865, he was fascinated by ornithology as a boy. This love led him to climb up and down steep faces and cliffs to seek out eggs and nests. The practice evidently stood him in very good stead, as his list of first climbing ascents in later life would take up the rest of this chapter. His name hovers over Scottish mountaineering even to this day, despite the fact he died almost a hundred years ago. From 1896 to the early 1920s he claimed fifteen of the thirty new routes on Ben Nevis *alone*. His ice axe is one of the two seals of office held by the Scottish Mountaineering Club president. That's how much of a big deal he was. Not content with Scottish hills, he summited the Matterhorn and led expeditions to the Caucasus, as well as being a member of a Himalayan trip. He died and was buried in Edinburgh, leaving a CV that earns him a place in the pantheon of world-class climbers, not just British.

As much as Raeburn was a man apart from his peers, the same case could justifiably be put forward for Creag Meagaidh.[14] In height and scale it is similar to Geal-Chàrn, but it is more complex. A myriad of ridges and corries emanate from the large summit plateau. It presents serious navigational difficulties in poor weather. The hill rises to an impressive 3,700 feet and is marked by a small cairn at the top. Many walkers confuse the summit cairn with the very, very odd, far larger one which sits just a hundred metres or so away. Like the summit cairn, this larger structure is man-made. It goes by the strange name of 'Mad Meg's cairn', and it has an apparently peculiar heritage.

Mad Meg's cairn is reputedly the final resting place of a woman who committed suicide sometime in the last few hundred years. The church would not allow her family to bury the poor soul, on account of her

---

14   Pronounced as 'craig MEHgay', meaning 'the rock (or crag) of the boggy place'.

taking her own life; consecrated ground could not be troubled by such an act. The family, therefore, decided to lay her on the highest ground they could, literally on the other side of the parish boundary (itself marked by Creag Meagaidh's summit cairn) just to spite the landowners. How much truth there is in this fable is unclear, but parts of it make sense. The cairn is a considerable structural undertaking in a very out-of-the-way place, so its construction would not have been an easy or frivolous endeavour. Its very close proximity to the parish boundary is also telling. Maybe one day it will be excavated, but I sincerely hope not. Leave poor Meg to the elements, I say. It is a wonderful, albeit tragic, tale that deserves to maintain its mystique.

I digress.

Of the many corries and ridges, the finest on the hill by far is Coire Ardair. Many consider it to be one of the greatest Highland corries, able to stand on a podium with Beinn Eighe's Coire Mhic Fhearchair and Lord Byron's favourite, Lochnagar. Having seen all three in the flesh I feel Coire Ardair is the finest. Its cliffs of the Post Face are simply magnificent, rising over 1,000 feet from the lochan below. If winter conditions allow, most climbers will tell you there are few places outside Ben Nevis where their sport is better.

But it is not the cliffs that are the object of desire here for the snow hunter. As usual, it is the gullies that concern us. The two main ones are Easy Gully and Raeburn's Gully. The latter is named after Harold, who completed its first winter ascent in the late nineteenth century. At the very top of this gully there lies a medium-sized, but deep, bowl. This place is highly unusual in that – like the Aisre Cham at Geal-Chàrn – it is one of the few places where snow has survived from year to year outside the Cairngorms or Nevis range (it did so in 2014 and 2015).

Life is confined in this hollow to only the most primitive types of mosses. This is in total contrast to the snow-free plateau just a matter of metres from it, where grasses and small plants thrive. In our snowy hollow the growing season is just too short for these. Around the fringes of the snow, out of the way of hungry sheep and deer, alpine plants grow. Of these the most photogenic is the starry saxifrage. This charming flower has five leaves, all brilliant white except for two yellow dots. In the middle, the stigma is a large violet ball from which emanate a number of white filaments, all topped with pink anthers. Google it.

The main gully on Creag Meagaidh is huge. It runs in a crescent-moon shape from just above Coire Ardair's small loch right to the summit plateau: the best part of a thousand feet of a climb, but longer as the fox runs. It is yet another gully with a misnomer. Easy Gully is anything but. It may well be in winter when there is ample deep, even snow to utilise, but ascend it in summer and be prepared for an experience like no other. The ground here is constantly on the move. In 2014, as I ascended the middle section, the whole floor that I was standing on started to slide backwards. It was terrifying, even if easily negotiated. In the past I've been in gullies where the ground was unstable, but never had I experienced a whole *section* of gully sliding away.

Easy Gully, however, is a treasure trove in summer. Apart from all the booty that can be obtained (ice axes, ropes, etc.), the gully is home to massive and weird snowdrifts. These are shaped by the unceasing drip-drip of a thousand springs above on the cliff faces. They are also the reason the gully is so mobile: the stones are lubricated constantly. The snow gathers there after funnelling down from the top of the gully. Avalanches tumbling down the Post Face also add to the great depth, ensuring snow is present right through to August and sometimes beyond. The shapes and scale of the drifts that remain here in late August are worth the walk on their own.

It was in Easy Gully that I managed to replicate a celebrated photograph from September 1899. The picture appeared in the Scottish Mountaineering Club Journal from 1901 and has become something of an iconic image. Taken high in the gully by SMC member Donald Cameron-Swan, it shows a behatted friend of his, Dr Kenneth Campbell of Laggan Bridge, standing beside a huge snow-bridge that spans the full width of the gully. Dr Campbell stands on the left-hand side of the picture and is dwarfed by the left pillar of snow which must be all of twenty feet high. The arch, according to Cameron-Swan, had a span of 'thirty-five feet'. In the foreground of the photo lie the remains of a previous iteration of the arch, now collapsed due to melting. It is altogether a marvellous image, of historical value. It is the earliest photo of a snow patch (or tunnel) at close quarters I am aware of in the UK.

The replication of this photo was in August 2016. Along with another hillwalker, Ben, I set off to see if such a thing would be possible. Ben was looking for interesting stories for a monthly column he wrote for a website. I had long fancied seeking out the precise spot on the hill and had a vague idea of where the photograph was taken. But, given that anyone who might have had knowledge of it had been dead for decades, it was a bit of a crapshoot. Right away, though, when we entered the gully – laminated 1899 photo in hand – it became clear from the skyline in the background we were definitely on the right track. As we climbed higher up the gully we marvelled at how (and why) those walkers from over a hundred years ago carried such heavy photography gear up the backwards-moving defile. We soon hit upon a likely spot for their photo. Luckily for us there were also quite a few drifts of old snow present, which added to the overall authenticity. Then, eureka, we saw a rock feature on the right-hand side of the cliff that matched up exactly with the one in the 1899 photograph. The skyline was perfectly aligned too. We had it. Ben stood where Cameron-Swan did and I where Dr Campbell was. It was as close to a perfect match as one could hope for, and Ben had his story.

* * *

Upper Tweeddale
(snow locations: NT180268, NT170250 and NT151241)

As the years roll on and on, I find myself becoming less and less well disposed towards sheep. Years ago, when I knew far less than I do now, I didn't mind them bumbling around on the hill, eating grass. They seemed like inoffensive, placid and daft creatures. In spring the fields were alive with the calls and sights of little lambs, gambolling across fields which were resplendent in their first flush of growth. What could be more natural? Those days, I'm afraid, are gone. Sheep (and deer) eat absolutely everything in their paths that has the potential to grow into a tree. In Scotland we still have far too few trees and shrubs, and a lot of this is down to the fact that our woolly and antlered friends munch them as they emerge from the ground.

The destruction of large tracts of the upland environment in Scotland lies squarely at the doorstep of mankind. Whether it was the killing of the last predator of the red deer (the wolf), the current over-management of heather moor (the better for people to shoot grouse) or overgrazing due to too many sheep, in terms of flora and fauna, our glens and hills are a shadow of what they once were. Once again, space restrictions – and self-imposed ordinances – prohibit me from delving too deeply into these subjects. There are plenty of well-written books and scientific papers that cover the topics better than I could, especially on the matter of grouse-moor management.

What do I have against sheep? Nothing, *per se*. It's not their fault they do what they do. It's just that there are so *many* of them all over the country that some parts of it feel almost monocultural. The Borders' hills – Upper Tweeddale in particular – are some of the worst affected, as far as I can tell. The wild, sparsely populated and unfrequented hills to the south of Peebles, around the farm of Manorhead, are real jewels. The unproductive soils of the hills would rebound with flora if there were a drastic reduction in the numbers of things roaming around that want to eat them. One only has to look at the trees and plants that grow in the inaccessible streambed shelves to see what *could* be. So it is, then, that whenever I walk in this fine part of the country looking for snow I do so with a tinge of disappointment that it should be so denuded of the life just waiting to erupt were it given the chance.

The three hills I visit most in this area are Dollar Law, Broad Law and Cramalt Craig. The first of the three doesn't hold the longest-lasting patches, but on its eastern slope, just to the east of Fifescar Knowe, a series of deep, north-east-facing gullies fill up readily in a winter of southwesterly snowstorms. This results in some excellent shapes and depths when visited on a warm April afternoon. Until they melt, these snows birth the splendidly titled Ugly Grain burn, which eventually gets gobbled up by the Tweed, miles to the north. Manorhead Farm, which the Ugly Grain passes, is extremely isolated. At well over 1,000 feet above sea level, it has the feel of a Highland bothy.

Broad Law, the Southern Uplands' second highest hill, carries a lot of strange paraphernalia on its summit. A radio beacon station, which looks for all the world like a lunar module, aids transatlantic planes in their navigation. The beacon is one of a global network of VHF Omnidirectional Radio Range (or VOR) stations set up as a radio navigation system for aircraft. Nearby a mobile phone mast adds to the incongruity of the scene. But as interesting as these things are, the real star of the show on Broad Law is its large northern corrie, Polmood Craig. This large scoop is Highland in feel, one of the very few in the region that can lay such a claim. Along its curved edge in spring forms the most impressive snow wreath in the Southern Highlands. Mini-crevasses appear, and it is quite possible to descend into them and be hidden from view to all watching above. The walk to this location is best from Cramalt Craig to the east. From there the finest views of the snow are seen.

Speaking of Cramalt Craig, this is the last in the Borders' trio that is worth a visit. (On a fine spring day, assuming one starts early enough and packs enough sandwiches, it is quite possible to take the snow from all three hills in.) I have yet to visit this hill in spring and not come across avalanche debris below its broad north-east shoulder. The avalanches occur firstly, of course, because of the volume of snow that accumulates on this high hill. Crucially, however, there is an abundance of well-lubricated grass. Any increase in temperature above freezing will see the bottom of the snowpack lose its grip. Eventually a critical mass is achieved and from there the outcome is inevitable: the snow lets go and down it falls. Large chunks of turf and earth are frequently to be seen lying churned up beside a car-sized lump of snow. It is certainly a place to avoid in the winter. A spring visit, though, is very pleasant – assuming the rain and wind stay away. On this exposed, flat ridge it is a very unpleasant place to be when walking into the teeth of a wet gale.

* * *

An Stùc (snow location: NN637432)

The story goes that, in 1878, a large group of about twenty men was led up to the summit of Ben Lawers, the highest hill in the Southern Highlands, by a chap called Malcolm Ferguson. They spent the whole day erecting as best they could a twenty-foot-high cairn. The reason for this eccentric endeavour was that they wished to bring their beloved local hill above the magical 4,000-feet barrier. It was previously thought to be this height, but when measured with specialist equipment by cartographers it was found to be lacking by sixteen feet. Incensed and highly motivated, Ferguson – like the Grand Old Duke of York – led the men up the hill then back down again. Alas for Ferguson and his workers, the Ordnance Survey ignored the man-made cairn, which has blown away in the intervening years.

Ben Lawers itself is the chief hill in a very fine grouping that encircles the northern part of Loch Tay. From Meall Greigh ('the hill of the horse stud') at the range's eastern terminus, the peaks stretch westward until reaching the summit of Ben Lawers, at an impressive 3,984 feet.

The journey into the foot of Meall Greigh, where one normally starts when completing a round of these hills, takes the walker up alongside the Lawers Burn (the derivation of Lawers as a name is disputed: some say it means 'loud water', while others say it translates as 'hoof' or similar). Beside the Lawers Burn there are over thirty shielings scattered on well-drained ground. A shieling is a summerhouse that was often used by local folk when they grazed their animals at higher level. They are very small and cramped and would have been, I'm sure, unpleasant in prolonged periods of poor weather. It is not known how old these structures are, but it's thought they date back at least several centuries.

Despite Ben Lawers' great height, it is not on the hill that the longest-lasting patch of snow in the range is to be found. That honour lies with its easterly neighbour, An Stùc.[15] Sounding little like how it's written, An Stùc is a superb peak and is translated into English as 'the peak'. Nothing could be more descriptive.

---

15   Pronounced as 'an STOOCHk', with the 'ch' as in *loch*, with a 'k' tacked on at the end.

When viewed from the east, on the summit of Meall Garbh – 'the rough hill' – An Stùc looks majestic. It is shaped like a huge pyramid, with even slopes tapering to an obvious point at the top. Its east face from here looks insurmountable to a walker. Indeed, care is needed when ascending this face, especially in winter. The rock is loose in places (as I know only too well), even if the climb up itself isn't nearly as hard as it looks from afar.

The long-lying snow on An Stùc tends to last until June or July, or exceptionally (à la 2015) August. It is situated in a pronounced hollow on the northern spur of the hill, where it gathers to itself lots of wind-blown snow from around the summit crags. Because of Ben Lawers' relative ease of access and popularity among walkers, photographs of the range are in no short supply in summer. Many hillwalkers become unconscious snow researchers when I gatecrash their social media posts. An Stùc's shapeliness ensures the snow photos keep coming. Consequently, it is yet another of the bellwether or signpost patches that serve as a reliable comparison year on year.

Alert readers will have by now realised that most of the snow patches I visit are out of the way, even if they're on accessible hills. An Stùc is no different. If the basin in which the snow sits is full, which it tends to be even in early summer, then one is obliged to creep very gingerly around its southern rim. The grass surrounding the snow is exceedingly slippery in wet weather. So much so that on my last visit, in July, I used crampons on my boots to get purchase on the grass beside the snow. Never before in my years of doing this had I used crampons on *grass*, only to take them off when reaching the *snow*.

That same day I was fortunate enough to encounter some of the rare and specialist alpine plants that Ben Lawers and its cohorts are renowned for. Among others, I came across a small clump of *Myosotis alpestris*, known charmingly in English as the 'alpine forget-me-not'; a stunning violet-petalled flower. To cap it all off, on the walk back out I was treated to a fine display of soaring from a golden eagle. The perfect way to conclude a memorable day.

nine

South of the wall

There is a curious school of thought in sections of polite Scottish society that all Britain's finest hills are located north of the border with England. This inexplicable theory can only be explained by a lack of familiarity. Many walkers in Scotland would, for reasons that I have not been able to properly ascertain, scarcely think to venture south. Some do, of course, but the vast majority do not. I am by no means a veteran of walking in England and Wales, but I have ambled on, I think, just about every upland part of England and quite a few in Wales. In my experience some of these areas are just as fine as parts of Scotland. Moreover, if we are talking about individual hills then very many indeed in Scotland lack the character and challenges of the best that Wales and England have to offer.

They have little to do with snow, but one of my favourite groups of hills in the whole of Britain is an almost unknown range in South West England called the Quantock Hills. These diminutive peaks rise to about 1,250 feet and are very much overshadowed by their neighbours on Exmoor. Few know that they were England's first Area of Outstanding Natural Beauty, having been designated as such in 1956. The main route to this range for most people is not done in a car or on foot: it is literary. Samuel Taylor Coleridge's most famous works (*The Rime of the Ancient*

Mariner and *Kubla Khan*, amongst others) were completed while the poet lived in the picturesque village of Nether Stowey, at the very foot of the Quantocks. In August the entire top half of the range is covered in the most beautiful carpet of purple heather that I have seen in Britain. One of the few truly wild red-deer populations of old inhabits the serene, wooded dells on the west side of the hills. It is one of the finest places I can think of.

The rest of the chapter is given over to some of the marvellous hills and ranges of England and Wales where snow can lie, or has lain, long. I go to some of these peaks regularly as part of my research for the annual snow-patch paper which appears in the Royal Meteorological Society journal *Weather*.

* * *

The Cheviot

Journeying south from the border between Scotland and England, the first stop is less than a mile from it. The Cheviot is a landmark hill in the Borders, conspicuous for miles from many vantage points. It has always attracted a high level of interest from travellers, and so – apparently – have its snows:

> *Sir* – I have read your paragraph about 'Snow Conditions on Cheviot.' It might be of interest to mention that 'pre-war I have taken parties up the Bizzle and struggled through snow pockets of about six foot depth during the early part of July.' These snow pockets or what used to be called 'Snow Flowers' frequently remain in deep drifts until late summer.
>
> – B.A. Parkes (ex-Lone Wolf),
> Berwick-upon-Tweed, June 14th 1951.[1]

This letter, written to the editor of the *Berwickshire News*, was the first in a series penned in response to a newspaper contributor mentioning seeing

---

1   *Berwickshire News*, 26 June 1951.

patches of snow on the Cheviot in early June and remarking upon how late it was. Mr Parkes's letter, followed by several thereafter by different correspondents (including one from Seton Gordon, remarkably), is a charming reminder that people have, for years, been intrigued by these long-lying drifts and moved enough to write to newspapers.

The Cheviot's name is ancient, probably Celtic. It may derive from the Brittonic word *ceμ*, meaning 'ridge', plus the suffix *-ed*.[2] At 2,673 feet it is the highest point on a large tract of upland Northumberland. The range in general gives its name to a breed of sheep, which is perhaps what most people will recognise the name for. (Someone formerly of my acquaintance used to call all sheep, irrespective of breed, 'stupid cheevs' in response to witnessing one of them doing something peculiar. The *cheevs* in question were Cheviots.)

The 'big Cheviot', as it is still known locally, has provoked writers down the years to gush in flowery, whimsical verse when describing even the most mundane of activities. One writer was so taken with the landscape and the simple-living people who dwelt in it that he felt compelled to write the following:

> In a word, a shepherd's life may be very pretty and poetical in Arcadia, but a Cheviot Tityrus[3] has other fish to fry, than piping like a bull-finch, or playing 'love among the roses' with some Daphne or Amaryllis, in full costume, as Fentum turns out his shepherdess for Jullien's *bal masqué*, with a straw hat, ribbons that would supply a recruiting party, crook in hand, and everything complete, barring the lamb.[4]

A reader could be excused for thinking, on the strength of the above extract, that the Cheviot is a spectacular hill of breathtaking loveliness. Though beauty rests in the eye of the beholder, I would not describe it as such. It is a large, mostly featureless lump with a very high percentage of

---

2   James, Alan G. (2019), *The Brittonic Language in the Old North*, Scottish Place-Name Society.

3   The shepherd in Virgil's *Eclogues*.

4   Maxwell, W.H. (1852), *Border Tales and Legends of the Cheviots*, London: Richard Bentley.

its slopes given over to heather and peat. Its main selling point is the spectacular views from the summit and the fact it lies right on the Scotland–England border. To the north-west the fertile verdure of the Scottish Border farms stretches out like an Astroturf carpet towards the escarpment of the Southern Uplands. On a clear day it is rumoured one can see Ben Macdui on the horizon, some 124 miles distant. I find this fanciful and have never seen it myself, but respected observers assure me it's gospel.

The Cheviot does possess a few parts which are worthy of note nonetheless, namely College Burn, Bizzle Burn and Bellyside Burn. College Burn faces south-west; the latter duo north. Of the three it is usual for the two north-facing ones to hold snow longer than College Burn, due to the preponderance of snow to blow from the south-west in winter. However, and this is a big 'however', if conditions allow, the College Burn area can have deposits of snow which make it difficult to believe one is in England.

In the aftermath of the so-called 'Beast from the East' snowstorm, which hit the country on 1 March 2018, colossal amounts of snow were dropped all over the country. At the College Burn, upwards of thirty feet was dumped. During a visit there over two months later I was able to walk among the small crevasses that had formed, with the snow still towering over my head. What a spectacle it was, especially given the delicious warmth that the high spring sun was giving off. It recalled to me the adventures of noted ornithologist Abel Chapman, who went to the Cheviot in May 1915:

On 14th May we had climbed out some 2,400 feet – beyond the head of the Bezzil [Bizzle] ravine – when further progress appeared absolutely barred by upstanding ridges of green ice that resembled nothing so much as a glacier … The sheer face of green ice exposed stood seven feet in vertical height … scarcely credible in the temperate zone of our British Isles, especially in mid-May … it remains to add that, even at midsummer, whole acres of snow still choked the Bezzil ravine.[5]

---

5   Chapman, Abel (1924), *The Borders and Beyond*, Gurney & Jackson.

Top Emerging into the light after ascending a snow tunnel. The roof of this part of the tunnel collapsed some days previously, leaving large chunks to melt on the relatively warm rock. © *Alistair Todd.*

Above Mud, soil and rocks cover snow debris deposited in a huge avalanche.

Top A runner stands beside the relic of the previous winter on Ben Macdui in November 2020.
The old snow was buried the following day, only to emerge once more at some point the following year.

Above This old patch of snow on Ben Macdui lies partially buried by fresh falls in November 2020.
The scalloped shapes are called 'ablation hollows' and are caused by the passage of warm air across
the surface. These edges attract dirt and debris.

Top A classic view of old snow patches clinging on to the cliffs. Here, at Aonach Mòr in October 2014, the old patches survived until the new snows of the following winter.

Above The author leans against the shear in a huge patch of snow at Aonach Mòr in September 2020. The shear was created by the patch to the author's right sliding down the granite slope, which had been lubricated by meltwater.

Ablation hollows and mesmeric colours at Ben Nevis's Point 5 Gully.
The vista here was described by one commentator as akin to 'the eye of a dragon'.

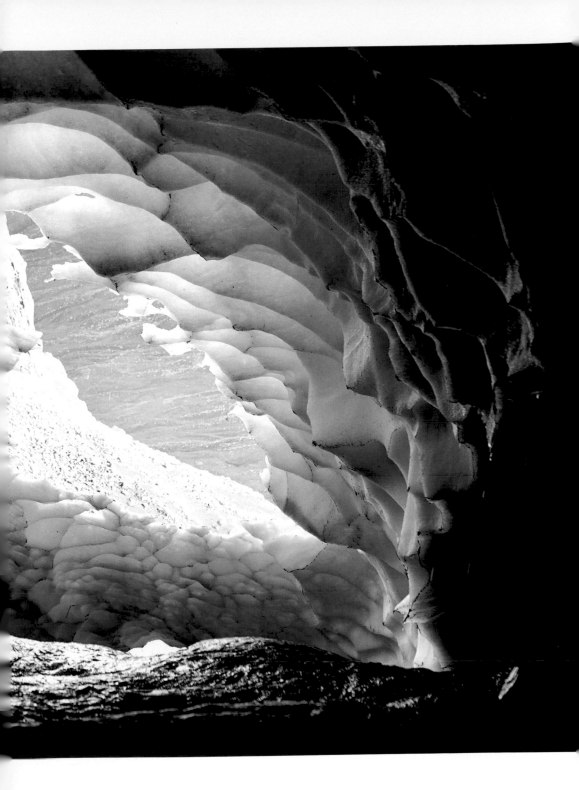

Top Three volunteers assist with the measurement of a massive relic of the previous winter, high on the north-east face of Ben Nevis in August 2015.

Above Huge drifts of snow remain at Hen Hole on The Cheviot in northern England in May 2018. This snow was deposited almost entirely during the so-called 'Beast from the East' in early March of the same year.

Top Marching towards the snow, high on a Cairngorms hill. © *Murdo MacLeod*.

Above A bitterly cold winter camp on Liathach in Torridon, February 2017. The wind chill was extreme, but the views that evening were some of the most stunning the author has ever encountered.

Top BBC commentator Andrew Cotter with Mabel and Olive, during one of our many walks.

Above Andrew Cotter again, this time on skis with his two dogs.

But the Beast from the East's crevasses melted quickly in the strong sunshine of May 2018, to the extent that they – and the snow they occupied – had vanished by the first week in June. Nearby, on 9 June, an observer went to the Bizzle Burn, just as Abel Chapman had done over a hundred years before, and saw the hill's very last snow. As he ate his sandwiches he noted how the small, dirty relic was the size of a serving platter. It was to succumb to the elements the same day. The observer wrote in a note to me the next day that he found this 'sad'.

* * *

Cross Fell, North Pennines

A mountain that is generally ten months buried in snow and eleven in clouds cannot fail exciting the attention and curiosity of a traveller.[6]

Outside of the Lake District, no hill in England is higher than Cross Fell. At 2,929 feet it commands panoramic views across the fertile Eden valley to the Lake District hills. It is also the highest point in the Pennines, which – along with the Grampians of Scotland – is one of the longest hill ranges in Britain. Anyone who travels to or from Scotland either by car or train has a superb, albeit distant, view of the Cross Fell tops on the east side of the road and railway track. Its subsidiary top, Great Dunn Fell, has a very obvious white golf ball structure on the top which serves as a radar station for the National Air Traffic Services. A necklace of snow wraps around Cross Fell in spring if favourable conditions prevail.

Cross Fell was once called Fiend's Fell. It is unclear exactly as to when, or why, the name changed. A local legend has it that an exorcism of said fiend (credited to St Augustine) was carried out and a cross planted on the summit, giving it a more wholesome aura and its new name. The so-called fiend that St Augustine exorcised was evidently not too keen on travel, as he or she fled only five miles to the north-west, along the Pennine escarpment, to take up residence on another Fiend's Fell. In truth it is likely the original name Fiend's Fell is drawn from the

---

6   Hutchinson, W. (1794), *The History of the County of Cumberland [etc.]*, London: F. Jollie.

shrieking noise often heard on the summit when the famous Helm Wind is blowing.

Like the big Cheviot, Cross Fell's shape is conical rather than craggy. Whereas the former has some interesting geographical features, the latter has rather fewer. It does, though, have very good historical records of snow-related events – good enough to put it near the top of the best-documented hills in England. The snow researcher *par excellence* Gordon Manley gathered many accounts for this area. One of the most interesting was his comment that 'the indigenous use of skis by the lead miners in the Northern Pennines has died out since 1900', which suggests that by this date it was no longer practicable to use them, as it had been in the past.[7] He cautioned, though, that this in part may have been to do with the demise of the industry as much as the reduced snow cover. It is a delightful thought to imagine lead miners in Alston or some other North Pennines village setting off from their houses after breakfast on skis.

Snow in general on Cross Fell is to be found on its northern side where it has melted everywhere else. Crossfell Well, a high spring sitting at grid reference NY685347, just at the foot of the summit bluff, normally carries a good drift into May. Its relatively open aspect, however, means that it is prone to rapid melting.

Remarkably, and perhaps exceptionally in the last fifty years, the ghosts of the old lead miners were roused from their graves on 1 July 1979. On that day, members of the Carlisle Ski Club hiked up to a gully near the summit and skied a large patch that was still plainly visible from the M6 motorway below. A local paper, the *Cumberland and Westmorland Herald*, ran the story a few days later. Over a month and a half later, on 18 August, a farmer by the name of Mr Willan saw two tiny white dots at the same locality while working in his fields, and no doubt they melted a day or two afterwards. This is the latest I have ever heard of snow lying on an English hill in the current era. It is very hard to imagine that date being beaten in my lifetime, alas.

* * *

---

7   Manley, Gordon (1949), 'The Snowline in Britain', *Geografiska Annaler* Volume 31, 179–193.

The Lake District

It seems rather futile introducing the most famous range of hills in England. So many wonderful writers, poets and musicians have eulogised the Lakes over the years that perhaps every native-born Englishman or woman has the memory of them running through their veins.

No writers are more associated with the Lakes than Dorothy and William Wordsworth, both of whom were born and grew up in Cockermouth. The landscape had a deep impact on them and they wrote extensively on it. It is pleasing to know that these two most famous of Wordsworths were also snow observers.

> It was a Cove, a huge Recess,
> That keeps till June December's snow;
> A lofty Precipice in front,
> A silent Tarn below!
> Far in the bosom of Helvellyn,
> Remote from public Road or Dwelling,
> Pathway, or cultivated land;
> From trace of human foot or hand.[8]

It may not rank as William Wordsworth's best work, but it is highly instructive for the mention of June harbouring snow from the previous year. The 'silent tarn' is, without question, Red Tarn below the north-east-facing cliffs of Helvellyn. The 'cove' is likely to be Brown Cove, just north of the summit. It is almost always the place on the hill where snow lies latest. William's sister, Dorothy, in a publication of 1817, mentions that snow remained on Helvellyn above Red Tarn in late June.[9] Given that both the Wordsworths mention Helvellyn's snow in June, it is possible and, perhaps, likely that William's observation in 'Fidelity' was taken on the same day as his sister's, conceivably during a family excursion. Having sat at Red Tarn myself and looked up at the last chunks

---

8   Wordsworth, William (1815), 'Fidelity', *Poems in Two Volumes: Part Two*, London: Longman.
9   Manley, Gordon (1952), *Climate and the British Scene*, London: Collins.

of the previous winter's snow that were about to expire, it is heartening to think that I could have been sitting just yards away from where they picnicked and penned their verses.

Helvellyn itself is the finest hill that I have walked on in the Lakes. At the start of the chapter I mentioned that some English hills were more than equal to their Scottish counterparts: Helvellyn is just such an example. Though modest in height, at just 3,117 feet high, it is a complex hill of airy ridges and high cliffs. Striding Edge – a tentacle reaching out from its summit and stretching to the east – is the best ridge-walk in England, offering superb scrambling for those who are so inclined. Swirral Edge provides an excellent descent and horseshoe route back to the 'silent tarn'. The vista from Red Tarn on to Helvellyn's north-east face is one of the classic English mountain views.

On Helvellyn's Brown Cove nowadays, June snow is still possible to find (2010 was a recent example), though it is becoming increasingly rare. This is also true for nearby hills such as Great End, Scafell Pike and Esk Pike, which normally hold snow late into spring. Gordon Manley wrote in 1952 that all the above could expect to have snow till June in a normal year. 'Normal' nowadays means something different. Things have changed remarkably in a relatively short period of time. Like the August snow at Cross Fell in 1979, replicating Manley's observation in 1951, when Brown Cove held a small drift in mid-July, seems unlikely any time soon.

* * *

The Peak District

To Scots, the term 'down south' means everything from Carlisle to Southampton. This echoes how people from northern England playfully mock southerners for calling anything above the Watford Gap 'The North'. It has never been clear to me where the North begins. Leaving aside the Watford Gap stereotype, the question is vexed. The North York Moors, the Lake District, the Cheviot Hills: all these ranges are unequivocally in 'The North'.

So, then, where does the Peak District lie? It surely can't be in the

North as it forms the most southerly spur of the Pennines. It is *undeniably* not in the South. It is also too far north to be in the Midlands. It is, therefore, in no man's land. QED.

Whatever its location, the Peak District National Park is immensely popular, attracting over ten million visitors each year. This is hardly surprising, given its proximity to some the largest cities in England. Manchester, Leeds, Stoke-on-Trent and Sheffield all lie within striking distance of the Peak District's borders. Most of the ten-million-plus sightseers will not ascend the highest hills or even venture beyond the periphery of the roadside. Despite the throng, for those willing to seek it out, there is some wild country to be had.

Kinder Scout is the highest hill in an area that is known locally as the 'High Peak'. At 2,087 feet it doesn't trouble the list-makers of England's highest hills. Nevertheless, there is nothing higher in England farther south than Kinder Scout, not even on Dartmoor. It is, therefore, magnetic to walkers because of that. The same walkers, if they felt so inclined, could walk across its huge and impressive summit table for three miles and not once drop below 2,000 feet. One would be hard-pressed to do that in many places in England, even in the high Pennines. Kinder Scout is a totem in the area, much-loved and visited. It is also much observed and has been for centuries. Once again, snow observers in days gone by do us proud:

This year, 1615, January 16th, began the greatest snow-storm which ever fell upon earth within man's memory. It covered the earth fyve quarters [45 inches] deep upon the playne. It fell ten several tymes, and the last was the greatest, to the greate admiration and fear of all the land; it came from foure p'ts of the world, so that all e'ntries were full, yea, the south p'te as well as these mountaynes. It continued by daily by increasing until the 12th of March, (without the sight of any earth, eyther upon the hilles or valleyes) upon wch daye, being the Lord's daye, it began to decrease, and so by little and little consumed and wasted away till the eight and twentyth day of May, for then all the heapes or drifts of snow were consumed, except one upon Kinder Scout, wch lay till Wit-sun-week.[10]

In 1615, Whitsunday fell on 7 June. Old snow visible on any part of the Peak District in June nowadays would be unthinkable. It is just about conceivable that an Arctic blast during the first week of the month could bring a dusting to the range's very highest tops, but it is probable that such snow would be ephemeral. Nowadays one is lucky to find old snow there in April. There are occasional exceptions. One notable example was 3 April 2012. During the early morning, very cold air edged south-wards and an intense fall of snow resulted in road closures in the Peak District and Pennines. Three weeks later large drifts were still in evidence across Bleaklow and other high hills in the area. The very last to go that year clung by its fingertips until the first few days in May. Another unusual event happened in 2013. Weather fronts associated with a deep depression to the south-west of Britain collided with cold, easterly air to bring intense falls of snow to the Peak District from 22 to 24 March. That year, the last patch gave up the ghost around 10 May.

All the snow that I have talked about so far in the book has been above ground. This may sound like an obvious thing to state, but in the Peak District there is at least one example of a phenomenon that occurs nowhere else in Britain. This unique feature is a deep fissure called 'Eldon Hole'. This pothole, dubbed the 'fourth wonder of the Peake',[11] is little-known, except to speleologists[12] and locals. Entrance to the pothole cannot be achieved without the aid of specialist equipment, so con-sequently it is visited only rarely, even in summer.

With its entrance lying in a northwest–southeast alignment, and sitting at 420 metres above sea level, Eldon Hole is prone to collecting snow when winds blow in from a north-through-easterly direction. The nature of the pothole's internal structure means that any snow blowing

---

10   Hall, T. S (1863), 'Derbyshire (England) – description and travel' [the account was transcribed from the Youlgreave parish register by Hall].

11   Cotton, C. (1683), *The Wonders of the Peake*, London: J. Brome.

12   People who specialise in the study of caves.

into its opening is relatively unimpeded until it hits the floor, resulting – given the right conditions – in very deep accumulations. Moreover, because the cave is enclosed, offering complete protection from solar radiation and strong winds, melting is reduced dramatically when snow gathers here. If the accumulations are large enough, the snow will significantly reduce the ambient temperature in the pothole through spring and summer by a process of radiation, thereby further slowing the melting process.

In 2013, I was contacted by a caver who had had to abandon a trip to Eldon Hole during the first week in May because snow blocked the bottom cavern, and it was still present on 4 July when he returned. I noted this event in the *Weather* journal paper I wrote that year. Since that account, other evidence has come to light (personal communications from cavers). Firstly, deep snow was seen in June 2010 by S. Sharp, and the same drift – much diminished but still substantial – persisted in November, five months later. It is a reasonable inference that this snow endured to the first heavy falls of the season, which came only a few weeks thereafter. This is the only example we have in England where snow has almost certainly persisted from one winter to the next. Eldon Hole is a truly extraordinary location.

* * *

The Cotswolds
Nowadays, so little in the way of meaningful snow falls on the Cotswolds that no reports of long-lying patches have been communicated to me or anyone I know. The only reason I am including a section on them is because I live in hope, rather than expectation, that repeats of events past might happen once again.

The Cotswolds are associated more these days with weekend homes that well-off people from London return to after work on a Friday. But this is not a true reflection of the reality. Of its 800 square miles, eighty per cent of this designated Area of Outstanding Natural Beauty is given over to farmland. In a sleepy corner of this farmland lies the pretty village of Snowshill. Our old friend Gordon Manley said that this settlement, sitting at 800 feet above sea level, was 'likely to owe its name to past experience', which is hard to disagree with. Manley went on to say in the same book that after the extremely cold early months of 1947, a shady location of the northern Cotswolds held remnants of a drift until mid-July.[13] Now, were it not for the fact that it was Manley who reported this striking event, I would be inclined to dismiss it out of hand. Mid-July, on the Cotswolds? Surely Manley got that wrong? To my deep regret, Manley didn't say *where* in the Cotswolds this happened. One piece of historical evidence – actually, I'd go further, the *greatest* piece of snow-related historical evidence extant in Britain – casts a certain doubt in my mind about dismissing this, on the face of it, absurd claim.

During research in 2009 for a book I co-authored, I came across a reference to snow being visible in a Cotswolds quarry in August 1634. I thought initially that this was a ludicrous, highly overexaggerated claim. The report came from a book on the Cotswold Hills in Gloucester-shire, but the author did not cite the reference. However, Herbert Evans on a cycling tour gave a more precise description:

> So past Brockhampton Park, leaving to the left the quarries where in 1634 the snow and ice lay till August, past Sevenhampton.[14]

The large-scale OS map shows disused quarries that fit this description, just east of Brockhampton and at 850 feet altitude. The large expanses of smooth flat exposed ground to the north-east and east would be good gathering grounds for blowing snow. Hmm. Interesting, but still not conclusive.

Unfortunately, Evans did not give a reference either, but it did lead me to a more original and authoritative source in a book by William Dyde of Tewkesbury, published in 1790. On page 99 he wrote:

---

13   Manley, Gordon (1952), *Climate and the British Scene*, London: Collins.
14   Evans, Herbert (1905), *Highways and Byways in Oxford and the Cotswolds*, London: Macmillan and Co.

In January 1634, the greatest snow, that was ever remembered in the memory of man; and, it was attended with such extreme cold, violent, and tempestuous weather, that many people going from this market, were smothered and starved to death. And in the August following, great quantities of the same snow, and ice was to be seen in Brockhampton quarries, notwithstanding it was an extreme hot summer.[15]

Predictably, Dyde did not cite a reference either. I thought that, just maybe, he heard it as a child from an ancient relative who had, in turn, been told of it by their grandmother who had witnessed it first-hand. That turned out not to be the case. I came across some other bread-crumbs, just enough for me to conduct an online search among the dusty archives of the Gloucester Records Office. I found something that I thought likely to be the original source, so made an appointment to go in and view it.

With low expectations, I submitted the reference number to the helpful archivist at the records office that Saturday. It took about ten minutes for her to find it, but when it arrived and I leafed through to the relevant folio, my hair stood on end.

It was clear to me that Dyde had also read the very account I was now looking at, as the wording was almost identical to his own. In wonderful handwriting, inked on vellum, was a description from the early seventeenth century. The account, though difficult to transcribe in entirety, stated that:

on January did fall the greatest snow that was known in the memory of man ... and in August following a great quantity of the same snow and ice did remayne at Brockhampton quares ... yet it was a most extreme hott summer.[16]

15   Dyde, William (1790) *History and Antiquities of Tewkesbury from the Earliest Periods to the Present Time*, Tewkesbury: Dyde and Son.
16   *Account book of Giles Geast Charity*, Gloucester Records Office, reference D2688/1: folio 79v.

I was beside myself with excitement upon reading this. Of course, a handwritten entry in a book is proof of nothing except the person who wrote it. The question posed itself, though: *why* enter something like this if it weren't true? Why go to the trouble? Maybe, just *maybe*, given the right circumstances, some of this snow could have held on until August. It is such a tantalising story.

I mention this story at some length mainly because I think it's a tale worth telling, but also because it illustrates the value of accurate recording by the author, as well as perseverance and diligence by the researcher. In this game, these two attributes are not so much desirable as essential.

* * *

The Carneddau, Snowdonia, Wales

As far as this section of the book is concerned, some might contend that leaving North Wales till the end is keeping the best till last. There is certainly some merit in this position. Snowdonia's peaks are rightly lauded for their beauty and the rich variety of walking and climbing available. Tryfan is a pyramidical peak of sublime proportions, easily capable of being included in a photo-finish for the 'handsomest hill in Britain' award. Crib Goch is one of the classic British ridge walks and among the very best in terms of sport. All over this range soaring peaks and deep cwms leave the visitor awestruck at the range and depth of the landscape on offer.[17] Snowdon itself – or Yr Wyddfa to Welsh speakers – is the highest peak in Britain outside the Scottish Highlands and possesses a set of cliffs on its north-east face that many Scottish hills can only envy.

The most northerly section of Snowdonia encompasses a range called the Carneddau – literally 'the cairns' – in Welsh.[18] It has the largest collection in Wales of hills over 3,000 feet, including the one we are concerned with: Carnedd Llewelyn.

In the main, Wales's hills are too craggy and steep to hold on to snow for extended periods. (Snowdon's great height – 3,560 feet – would, if it

---

17   A cwm is the Welsh equivalent of a corrie.
18   There are three distinct sections to Snowdonia's hills: Snowdon, Glyderau and Carneddau.

were more rounded and had gullies rather than cliffs, allow significant accumulations of snow.) The Carneddau are different from the peaks to the south. Carnedd Llewelyn has a whaleback ridge running for over a mile north-east from its summit cone. Some decent-sized cliffs sit above cwms on all sides of the hill. Despite all these glacially carved features, the place on Carnedd Llewelyn that virtually *always* holds snow later than anywhere else in Wales, and often outlasts England's as well, is called Y Ffoes Ddyfn (pronounced, very approximately, as 'uh-FOICE-thuvin', with a 'th' sound as in 'the').[19]

Back in the early part of the twentieth century this location, translated as 'the deep cut', aroused some interest in a local man by the name of J.R. Gethin Jones. A native of those parts since birth, he used to wonder why the 'snow always remained latest on Carnedd Llewelyn' rather than on Snowdon. To resolve this matter he 'undertook a bit of local exploration'. In the resulting paper, which he wrote for the inestimable *British Rainfall* magazine in 1909, he gives a superb overview of the location of Y Ffoes Ddyfn and accompanies it with valuable pictures.[20]

With a theme that has not only been running through this chapter but the book itself when historical records are mentioned, the amount of snow present at 'the deep cut' back in Jones's day was significantly more than is present today. In the paper he described a visit in June 1910. He concluded that the last snow on Snowdon disappeared generally at the beginning of June, but in Y Ffoes Ddyfn at the beginning of July. During the very snowy winter of 2009–10, by comparison, the patch at Y Ffoes Ddyfn persisted until 28 June.

Even this late date by today's standards would appear to be earlier than one might have expected in the mid-nineteenth century, according to writer James Orchard Haliwell. In the late 1850s he walked with a guide on Carnedd Llewelyn and noted:

---

19   During the period 2010–2019, Wales's snow has been the last to vanish six times, compared to England's five.
20   Gethin Jones, J.R. (1909), 'The Spot in England and Wales Where Snow Lies Latest, with Observations of Snowfall on the Snowdonian Range', *British Rainfall*.

the cold was intense; I should suppose it to be five or six degrees below the freezing point, and this while the sun was shining at noon day ... The snow lay very deep in some situations. We walked over a ravine which my guide informed me was at least forty feet in depth. In this, and familiar places, he said that it would not disappear until the close of August.[21]

As with the account from the Cotswolds, we cannot take one anecdote as proof of anything except itself. Or, as Adam Watson and I state in our conclusion in *Cool Britannia*:

> Although much of the above historical evidence is anecdotal and qualitative, it appeared independently with many observers in different years and regions. Because of this, we infer that observations were unlikely to be atypical, such as being concentrated by chance in an unusually snowy year or area. Comparisons involving quantitative observations from recent decades add confidence to this inference. To sum up, it is reasonable to conclude that more snow lay on the hills of Britain during the 1700s to early 1900s than in decades since 1930. Accounts from the lowlands of Scotland and England fit with this.[22]

Today, because of accounts like this being more widely known, many observers keep keen eyes trained on Carnedd Llewelyn's Y Ffoes Ddyfn to see when the last vestige of snow melts. The pioneering onlookers of the olden days fuel the desire to keep these visits and observations going. I find it enormously gratifying that it is not just in Scotland that people are apt to record long-lying snow. It tells me that there is a thread that links like-minded people across Britain, a thread that grows stronger by the year.

---

21   Watson, Adam and Cameron, Iain (2010), *Cool Britannia*, Bath: Paragon Publishing.
22   Ibid.

ten

Modern perils

Although snow is one of the motivations for people like me going up into the hills, for others the need or desire to do so has remained virtually the same since people started doing it in large numbers some 100-odd years ago. For many it is the re-engagement with nature and the wild; for others it is the challenge. For some others again, it is a combination of all of these. The thing that has really changed in the last ten to fifteen years, however, is the leap forward in technological capability, especially the handheld device. Allied to this change in technology is the now almost ubiquitous use of social media and online applications. Before we delve into that vexed subject, complete with its many pros and cons, it might be useful to look at the modern attitudes to hillwalking in general.

The reasonably observant, societally mindful citizen will have noticed a change in their lifetime regarding risk and the perception of risk. I mean here the general risk, real or imagined, that one comes across via the workings of everyday life. Risk when walking or climbing is just an extension of this. The reader may remember from the opening chapters that my occupation revolves around the management of risk (in the construction industry). In addition to this I've been a Chartered Member of the Institution of Occupational Safety and Health (CMIOSH) for over ten years, so I am reasonably conversant with the subject.

There is little doubt in my mind that society, in general terms, is considerably more risk-averse than it was when I started hillwalking properly in the early 1990s. For many years now, as it seems to me, the public have been looking more and more to the State or some third-party body to provide the things that a generation ago they would have done – in large part at least – for themselves. The reasons for this are many and complex, but a significant one (perhaps *the* most significant in my eyes) was the changing of the law in the 1980s to allow individuals to litigate in an American 'no win, no fee' model. This move by the government, almost certainly to cut down on the legal aid bill, had unforeseen consequences.[1] One of them was the eye-watering expense to the public purse as a result of so-called 'ambulance chaser' solicitors scouring the country for plaintiffs. Thirty years on from these changes, the effects are clear to see (England and Wales figures):

There were 143,167 practising solicitors in 2018, up 2.7 per cent on the year and 27.3 per cent over the past decade. However, there were 50,684 in 1988, and growth was nearly 50 per cent in each of the following two decades.[2]

The knock-on effects of such a legal-system-heavy society (in relative terms) are everywhere. Daytime TV adverts for 'no win, no fee' law firms pay the salaries of the people making the programmes we watch. Children's play parks that were once concrete-floored are now rubber-floored, surreal environments where the swings are shorter and the climbing frames lower. (I am not saying this change is a good or bad thing, merely observing that it *has* changed.) Disclaimers are put on everything from sandwich wrappers to car park signage. They say effectively the same thing: 'The consumer uses these services at their own risk'. Safety legislation in the workplace is another area that has

---

1   Critics of this move would say that it was not at all 'unforeseen'.
2   Legal Futures. 2021. *Growth in number of solicitors starts to slow – Legal Futures*. Available at: www.legalfutures.co.uk/latest-news/growth-in-number-of-solicitors-starts-to-slow [accessed 13 May 2021].

grown massively since the late 1980s. This expansion is another of the reasons why the legal profession has grown to the bloated level it is now. We look to the law to solve problems that would once have been dealt with by the political process. The citizens of this country now place more faith in judges than they do in politicians. To my mind this is a damaging and unhealthy space to occupy.

Gradually, by osmosis, the general public has adopted this mindset as well. Whether we like to admit it or not, our collective societal toleration of risk has reduced and, consequently, created a void that we look towards others to fill. One example is the attitude of 'If I get lost I'll just phone Mountain Rescue'. I've seen that particular line written on internet forums a few times. In other words, the person appears to be happy to outsource their responsibility to some other poor soul. It is not an effective or societally egalitarian strategy to adopt. The thought of having to call Mountain Rescue fills me with horror. I'd sooner crawl off a hill than put someone else out.

This risk-aversion void is now occupied by a whole host of institutions and companies that barely used to exist: advice bodies, training centres, mountain guides, government-funded information institutions and websites. On and on it goes. None of these organisations or individuals have formal legal competence. But we, the hillgoing public, are expected to follow the advice they give to the letter.

The following list of equipment appears on a popular outdoors website and is deemed 'essential' for summer walking. The list is produced in full:

» Rucksack – about 35 litres
» Boots (with ankle support and soles which will grip on rock, grass and mud)
» Waterproof jacket (with hood)
» Waterproof overtrousers
» General trekking trousers (not jeans or cotton material)
» Thermal top
» Fleece top
» Gaiters

- » Warm hat
- » Gloves or mitts
- » Spare layer, e.g. fleece top
- » Compass
- » Map (waterproof or in waterproof case)
- » Watch
- » Torch (preferably a head torch)
- » Food and drink
- » Emergency survival bag (polythene is OK) and group shelter
- » Whistle
- » First aid kit (small)
- » Mobile phone

The organisation that produced this list is an excellent one. I have walked and worked with them on several occasions. Their members are knowledgeable and – I'm sure – issue advice to the hillwalking public that they feel is sensible and proportionate, no doubt after consultation with various other bodies and individuals. But let's look at that list in some detail. First, there's the cost. Even conservative calculations of the equipment leave little change from £700, and even that doesn't include a mobile phone, which most folk carry as a matter of course nowadays.[3] Multiply twice, assuming a couple want to walk together, and you're looking at something which is borderline prohibitively expensive to average would-be walkers. If I'm completely honest, of the twenty items on the list I routinely carry only half of them, with another four sometimes (normally in winter) and four never. If I have looked at the weather forecast, and seen that it's set fair and warm, I'll carry even fewer: maybe seven or eight. On a fine summer's evening I'll think nothing of walking to the top of my local hill Ben Cleuch carrying nothing on my back except a camera case.

---

3   A decent pair of boots alone will leave no change from £120. A waterproof jacket and pair of overtrousers will set the walker back another £150. Add a cheap pair of trekking trousers in and you're at £300 before fleeces, gaiters, hats, etc. These are all conservative estimates.

Second, the issue of practicality. If one has a settled anticyclonic weather system in summer – with no reasonable prospect of heavy rain or high winds – then taking waterproof jacket and trousers is, as far as I am concerned, an unnecessary burden. Doubtless my profession affects my personal thinking here but in risk management the phrase 'reasonably practicable' is one that is sacrosanct. All this phrase means is the computation of potential risk being weighed against other factors, such as cost.

The third issue I have with lists like this is the charge being levelled at me for being irresponsible should I choose not to take any of the items when I go on to the hill. I would be much happier if websites in general didn't stray into the realms of telling people what to do (i.e. something is *essential*). The term *essential* is subjective.

I remember quite clearly when jeans and trainers were considered acceptable hillwear. Photographs I've seen from the 1980s show kids, some as young as five or six years old, above 3,000 feet wearing wellies, trainers and sometimes sandals. Their 'waterproof' jackets offered only limited protection against wind and rain. Overtrousers, if they were carried at all, were of the same type. Despite this, many responsible parents took their kids up into the hills to try and instil into them the same love for the place that they themselves had. When I say 'responsible', what I mean is parents who were competent hillwalkers and who made sure they'd checked the weather forecast, brought plenty of food and water, and knew their way off the hill in a hurry if things took a turn for the worse. The responsible parent would also limit the scale of their ambition on a given day to the capabilities of the youngest child. It used to be a regular occurrence – as a total percentage of the hillwalking population – to see small children out. Can I be alone in wondering why so few now do it?

No sensible or reasonable person would take issue with a website that gives a handy list of things to look out for when going hillwalking. After all, what is a weather forecast but a guide as to what hills one might want to avoid on a given day? But I think it is unwise and overly paternalistic to advise the general public what is *essential*. If someone must be *told* that they need to carry a map and compass when venturing on to a remote hill the chances are they're not going to know how to use them. Therefore, the authority of self-appointed guardians of our hill safety must always be questioned, and never be allowed to become received wisdom. Attitudes can harden like concrete, so the mix must be stirred to keep it from setting.

Part of the issue, at least for me, is that I feel my personal autonomy is being challenged. I cleave to the dreadfully old-fashioned view that personal responsibility is the single most important attribute a person can possess. From it everything flows. But that is only part of it. For good or for ill, I have always had a contrarian mindset. If the flock flies right, my instinct has always been to fly left. Over the course of my life I think this attitude has helped me, not hindered.

Does anyone, really, like being told what to do?

* * *

In terms of the many specific changes to equipment and practices over the years, some have been welcome, but others less so. I'll speak about a few here.

The first one is the route card. This device is – as the name suggests – something that is filled in prior to going into the hills. At its most basic level it is no more than telling someone close to you where you intend to go and, roughly, what time you expect to get back. Some opt for a more formalised version and write it down. It is then given to a loved one or someone in a position of trust, the better to find you if you fail to return by the allotted time and end up lost or incapacitated. A route card can also be displayed through the windscreen of one's car.

I have two main objections to route cards as 'aids'. The first is that filling in such a thing ties a walker to a set journey, with little scope for deviation to a different glen or hill if the weather turns. I have frequently had to change plans when faced with poor or atrocious weather. By so doing, did I leave myself open to the charge of being irresponsible, or should I have just ploughed on against the weather because that's what I said I was going to do? Did deviating from my predetermined route place an unacceptable risk on those that would seek to rescue me? It also goes against the spirit of being outside. Or, as Mountaineering Scotland state on their webpage on the pros and cons of route cards:

> The imposition from outwith the mountaineering fraternity of an obligation to complete an over-elaborate route card for every expedition however mundane is not acceptable to many mountaineers. Such a requirement implies that hillgoers are in some way accountable to officialdom. This conflicts with the ethos of mountaineering.[4]

The second objection I have to them in the form of physical cards displayed in vehicles is derived from the tongue-in-cheek term *burglar's friend*. Advertising where you've gone and what time you are likely to return is a lip-smacking prospect for a potential miscreant. 'Dear burglar, you have at least two hours available to ransack my car, safe in the knowledge that I'm up on the hill and can do nothing about your criminal act.'

All that said, the *premise* of telling someone where you are going is not really mountain-craft, but good manners and common sense. If travelling somewhere solo I always inform a loved one where *roughly* I'm going, and I'm sure many folk like me do the same. What I am wary of in the future – and I can see a day when it happens – is that people will be obliged to inform a government-aligned or approved website where they intend to go if there's a chance they might need rescuing. Pressure will be put on the hillwalker that if they *don't* do this then their chances of being rescued – and thereby of surviving – will be reduced.

The best and most profound advance in technology over the last ten years when hillwalking and mountaineering, without question, is the ability now to navigate without error. Satellite-based navigation systems, such as the omnipresent Global Positioning System (GPS), have revolutionised navigation and are used by, if my experience is anything to go by, a good deal of the hillgoing population. Gone now is the need to keep an Ordnance Survey map around your neck in a large, transparent pouch.

---

4   www.mountaineering.scot/safety-and-skills/essential-skills/navigation/route-cards [accessed 13 May 2021].

(Having one of these constantly blowing in your face when you're walking into a stiff wind is a pain.) A mobile phone can carry the full 1:25,000 Ordnance Survey Landranger series in its memory and can pinpoint the holder to within feet of their location. If someone is on a plateau in thick mist and wishes to walk in the right direction, then all they need do is remove the phone from their pocket and the application will show them not only where they are, but what direction they are facing.

Some purists eschew using these devices on the grounds that it encourages laziness. Knowing how to use a map and compass, they say, is essential, and preferable to mobile technology. This position is broadly fair, but a tad curious. Technology makes life easier and – ultimately – more enjoyable. We no longer wear hobnail boots and tweed jackets, as we now have better waterproof materials and equipment. It is more lightweight. Does the purist who waves away the GPS also cling to his or her Victorian ice axe and walking breeches? Things move on. They evolve.

'What happens if the battery goes?' is frequently posited by way of a counter. To this question I answer that I always carry a small powerpack in my rucksack in case it is needed (although in ten years it has never emerged from my pack). In any case, my almost obsessive interest in maps over the years has allowed me to have a reasonable internal map in my head wherever I tend to be walking. Prior to going somewhere I've never been before, I have an idea of the hill and its topography based on looking at the Ordnance Survey maps. Personally, I cannot understand anyone who *wouldn't* do this. Maps are wonderful.

So good is the technology available currently that I will confess to not carrying paper maps and compasses on many of the trips I undertake. The brand of mobile device I have used over the last ten years has never once let me down on the few occasions that I have had to use it in anger.

To summarise: unless the current climate of risk aversion reverses or slows down then we will hear increasing clamour for hillwalkers, or people whose pastimes may require them to be rescued, to carry some form of insurance. In effect this will professionalise a cohort of people who currently do it voluntarily. To be fair to the various mountain rescue groups, part of the cohort to which I refer, they have always resisted the compulsory carrying of insurance, and pooh-poohed any notion of people being charged for their services. This is good, and I hope it continues. I fear, though, that the tide of public and political opinion which waxes will be irresistible. We shall see.

In terms of technology there is little doubt that some modern items are a step forward. Lightweight waterproof gear, boots and rucksacks make for a far more pleasant walk or trip compared to the old kit. As discussed, maps able to fit on the palm of your hand *and* which show you the direction you're going in are likely to limit mountain rescue call-outs.

* * *

A couple of years ago, I asked two friends a series of questions during a walk. What would they rather be without, their internet-enabled phones or their cars? The phone lost, perhaps predictably given the value people place on convenience and independence. What about a cooker? There was some debate about this, but in the end the cooker won out three to nil. A washing machine? This is where opinion started to differ. One of my friends stood adamant that the lack of washing machine was surmountable, while my other friend and I said not. (Perhaps tellingly, the friend who opted for the washing machine had two children.) What surprised me, when asked the same question about the following items, was the phone won either a majority or clean sweep: TV, chocolate, alcohol, books, laptop or computer.

Now, the three of us are all men in our mid-forties. If we were to ask men and women ten or even twenty years younger the same questions, would we get the same answers? Probably not. What point am I trying to make here? Mobile phones and their associated applications have become absolutely and utterly ubiquitous, to the extent now that many people wonder how on earth they'd manage without them. There's no need to list here their uses. Chances are you have one either in your pocket or lying close to you. From an outdoors perspective, as outlined earlier in this and other chapters, they fulfil a variety of uses, from navigation to camera, from notepad to weather forecasts.

When I am on the hill I very seldom, if ever, look at my phone, and only then to confirm a coordinate or use an application to determine a distant peak that I cannot identify by sight. It gives me peace of mind to know it's there if I should have need of it, but I try never to rely on it. In ten years of carrying it as a navigation aid I have seldom had cause to use it when extricating myself from genuine potential peril.

You can probably bring to mind, without too much trouble, news stories where a mountain rescue team has been called out to escort lost hillwalkers off some Munro, Lake District Fell, or Welsh 3,000er. Almost without exception the excuse for getting lost will be related to not having the correct navigational tool. In multiple cases I've seen and read about, the rescued person or party was at the mercy of Google or Apple Maps, thinking either of them sufficient. And this is the fulcrum of the problem. So dependent have people become on their phones that they are lulled into a false sense of security. 'If I have my phone I'll be all right' seems to be a prevalent attitude. But even the very best and most helpful pieces of equipment are only as good as the person to whom they are attached. The most up-to-date, super-flashy model of mobile phone is useless and worthless if it does not have the correct type and scaling of map on it, or if the user lacks the wit to operate it. Reliance on technology and technology alone is a single point of failure. This is indeed a modern peril.

* * *

As much as over-reliance on the mobile phone as a navigational tool is a modern peril, there are others. The one I have specifically in mind, however, is also a boon. I refer to the pervasive, inescapable spectre of social media.

From the perspective of research, social media – if used correctly – is one of the most useful tools currently available. At a talk in Aberdeen in late 2019 given by Cairngorms National Park Authority, I heard how their staff used an exercise-tracking application to work out where people were going in forests, the better to see how this could affect the mating habits of rare, ground-nesting birds: a really useful way of harnessing a technology in a way that the inventors could not possibly have known at the time. The architects of the social media sites I utilise probably *did* have this sort of thing in mind when they concocted it. The ones that provide the largest number of visuals (somewhat predictably, Facebook, Instagram and Twitter) are the ones that I frequent.

Contained within these three sites is a truly vast repository of photographs and information. To complement these there is usually a refreshingly large store of goodwill. It typically goes something like this: either someone has a day off work or it's the weekend. Intent on making the best of the good weather that is forecast they take a trip to a hill they've been meaning to do for ages. On the way up they'll take lots of photographs, usually on their phones. When they reach the summit, more photographs, and a video of the panorama. Sandwiches are consumed and the happy walker heads back down the hill to the car or whichever mode of transport awaits them at the bottom. Once home and showered, the walker uploads their photographs on to their social media platform of choice, which their friends and other members of the public will 'like' or comment on. This is where I come in.

At the end of a weekend of good weather in, say, late April, it is an absolute racing certainty that hundreds, if not thousands, of people will have ventured to the hills of Snowdonia or the Lakes and uploaded their photos. A very small percentage of these photographs will have displayed in them some of the last patches of snow visible, usually by accident. The trick for me is to find these photographs among the mass. It can be needle-in-a-haystack stuff. But all it takes is one good shot with a decent camera. Bullseye. This stranger's photograph has saved me the trouble and expense of having to go to the same place over the next few days. I will often comment on the photograph to see if the taker has a larger-scale version that he or she can forward to me so that I can better estimate the size of the snow that has almost gone unnoticed by them. Often pleased that someone has commented or taken an interest in their photograph, the owners of the snapshots are (usually) only too happy to send you what you ask for.

Because the work that I am engaged in has gained a larger following over the last few years, I often get unsolicited emails and messages via social media. Appearing on national TV and radio a few times has raised my profile, to the extent that when people do message me it's usually prefaced by words like, 'You're that guy who does the snow stuff' or 'You're that snow-patch guy.' Often the photographs they send are of no use at all. Occasionally, though, I'm sent a cracker; a genuine beauty of a patch that I either thought had vanished or didn't know was there in the first place, though these are quite rare. Whatever the quality of the photograph that is sent, I always thank the senders, who were good enough to go out of their way to send me the fruits of their labours. This is where social media is a terrific and valuable tool. The sharing and dissemination of certain information and pictures among the greater public is highly useful for research.

For every positive, alas, there is a negative.

* * *

23 August 2018

I should have known better. The internet – social media to be exact – doesn't do nuance or self-effacement very well. I (re)discovered this to my cost this evening when, on a popular Facebook outdoor page, I posted a provocative piece on not carrying certain equipment when hillwalking.

Current orthodoxy dictates that venturing above 2,000 feet in the UK necessitates getting kitted out for just about every conceivable eventuality. Things we are exhorted to consider are, amongst others, health, hygiene, first aid, weather, navigation, food, water, having a guide, terrain, footwear, clothing, route cards, map, compass, etc. *ad infinitum*. Not having or doing some of these things is, apparently, tantamount to a death wish, and is placing the walker (and the potential rescuer) in great peril. While there is absolutely no doubt that some of these things are valuable some of the time, the idea that all these things are needed all of the time is a nonsense.

In the post, I suggested a competent walker, familiar with the hill and terrain he or she is on, has absolutely no need for a map and compass.

Moreover, a waterproof jacket and trousers are also unnecessary in summer if you have a settled anticyclonic (i.e. a big high-pressure) weather system. I finished the post by adding that, on top of all this, the hills I had ascended the previous day (Aonach Mòr and Aonach Beag, both above 4,000 feet) were shrouded in thick mist when I climbed them, and I had little visibility. The level of vitriol in the responses was breathtaking.

Despite me pointing out that I was familiar with these hills to the point of saturation and knew exactly where I was going, regardless of the mist, I was classified as reckless, having a death wish, a fanny, arrogant, akin to a camper who leaves his rubbish after breaking camp(!), and various other forms of nasty personal abuse. No matter to what degree I was reasonable or considered with my responses, the bile continued un-abated, to the point that it was impossible to reply to everyone. (At all times I kept in mind Lao Tzu's adage of *responding intelligently to un-intelligent treatment*.) The moderators, not content that I was getting a proper pasting, weighed in and deleted the post, handing me a twenty-four-hour ban into the bargain.

So, what does this tell us?

In some ways it is indicative of the current poverty in general discourse. Intolerance is everywhere and immediate. My post was a reasonable one, albeit provocative and probably controversial. So what? I would have thought that is what online forums are there to cater for. Ninety-five per cent of the contributors, as well as the moderators, thought differently. They clearly imagined it set a worrying precedent. Someone with an alternative and dangerous point of view. One which didn't chime with the current tenet of safety above all else, even if it is to the detriment of one's enjoyment. My views were to be silenced and expunged.

Having the correct equipment for the conditions is essential if you are to go alone on a long day. But being in possession of all the things that you are told you need is completely unnecessary. Take, for example, a map and compass. I was told by virtually everyone in the long list of replies that these were required pieces of equipment. 'Why? What use is there in carrying something that you will never take out of your bag, because there is no need?' I asked. Among the many replies came banal statements like 'You might get disorientated' and 'You could get hit by a rockfall and the landscape changes'. To the latter suggestion I commented that if I were to be hit by a rockfall then not having a map and compass would be the least of my worries.

Once upon a time an opinion like mine would have been commonplace. Before the era of advice bodies, interest groups, mountain rescue teams and overzealous health and safety fanatics, people could be trusted to make their own decisions based upon an apprenticeship of going to the hill. They would build their knowledge slowly and surely, graduating to longer days and higher summits. You knew that there was no realistic prospect of being rescued, therefore you had to be more circumspect with your ambition. Those days are long gone. Nowadays, in an era of instant gratification, those long-held conventions have been bypassed. Too many people are rushing out ill-equipped and inexperienced. Mountain rescue call-outs are ubiquitous. The net result is that every social media poster seems to be a finger-wagging paternalist, chiding their naughty offspring for daring to not follow their advice to the letter.

Too many people, unthinkingly, swallow everything they are told when it comes to what one must do and what one must not do. I, for one, won't have it. If I think the advice I've been given is unnecessary or over the top, then I am absolutely within my rights not to follow it.

What I find most disconcerting of all, though, is the way in which debate, however reasonable it may be, is getting more and more difficult if you have an alternative point of view. Any casual observer of the political scene will know what I'm talking about. If someone disagrees with what you say they will shout and harangue you. They will resort to abuse and to invective. Over the last five to ten years there has been a noticeable breakdown in civility in this arena. There are many reasons for this, which are too long and too numerous to outline here. Suffice it to say that I find it depressing.

I have removed myself from that particular online group, incidentally, and all the online scrappers. It is impossible to wrestle with chimney sweeps and stay clean, so I shan't bother.

* * *

In the eighteen months since I wrote the above piece the situation has not changed. What *has* changed is that I no longer engage in any form of online dialogue. It is just about pointless. As the Argentinian writer Jorge Luis Borges said about Argentina and the UK fighting over control of the Falkland Islands, it is 'a fight between two bald men over a comb'.

On balance, however, social media is – despite its undoubted flaws – an immensely rich and useful seam of information. It often removes the labour- and energy-intensive process of going to a particular hill – energies and attention that can be diverted elsewhere. It is a medium that is here to stay, whether we like it or not.

eleven

Alone again (naturally): meditations on the outdoors

Solvitur ambulando. [It is solved by walking.][1]

'Why isn't it moving?' I say to myself. I crouch down and try to make myself look smaller, less intimidating. Walking slowly and deliberately towards the ball of feathers that I can see no more than fifteen feet away from me, I puzzle at the situation. Some of the feathers are white, but most of them are grey. It's definitely a ptarmigan. If I can just sneak up a bit closer to it I can get a good picture. 'I wonder why it isn't moving,' I say to myself, again. About six feet away now I look at where its head should be. There's no sign of it anywhere. Is it looking the opposite way? Wait a minute. I straighten my back and stand upright, directly over the bird. The reason I can't see its head – and the reason it isn't moving – is because it doesn't have one. Ah. Peregrine. Must've been a peregrine. An eagle or fox would've taken the whole thing. A peregrine ripped the head off and flew away. A tasty treat to finish breakfast, maybe.

It's a cold wind up at 3,700 feet. The T-shirt on the walk in seems a long time ago. At least the sun is out, though. It takes the worst of the chill out

---

[1]   Popular theory has this quote credited to St Augustine, though its origins may be much older.

the air. Below, at the bottom of the glen, the wind is whipping up white horses on the loch. I shall be glad to have that on my back walking home.

I pick my way carefully down the loose, steep slope. The rocks are as well bonded to the ground as my superglue repairs tend to be (i.e. not very), so care is needed. At this altitude it is still much too high for heather, whose root systems help knit the soil together and bond it. The vegetation here is a light brown grass that is about as hardy as it gets. I'm making good progress down it to about 3,000 feet, whereupon I hit the heather, which starts patchy and sparse but gets deeper with every few downward steps I take. I place my foot near to where I guess the ground will be best able to support it without turning an ankle. I repeat the process hundreds – thousands – of times descending from the plateau. You get used to it. In truth I enjoy it. The idea of it is worse than the reality. At least to me it is. The key is to keep the joints loose but poised. An unseen stone can roll your ankle any which way. A few of these can be anticipated on any given walk. It's unavoidable.

Every now and again, usually when I'm within about six feet or so of it, a brilliantly camouflaged grouse springs up from the heather like a ghost in a horror film, and it almost causes me heart failure. This, too, can be expected as a matter of certainty. It doesn't make it any less of a fright.

The mat grass that provides such a verdant hue to the stream courses in summer has changed colour to the yellowy-brown shade of autumn. There are many of these silvery water veins cutting down the hillside, creating a natural block between great swathes of heather. Nature's firebreaks. Heather itself does not like waterlogged ground. It is true to say that, as a general rule, if one walks on heather, one's feet will remain largely dry. The pay-off is the risk of turning an ankle. But heather also dislikes snow. If snow lies in one location on a hillside for any longer than a few months unbroken, year after year, the heather will struggle to take hold.

Anyway.

As I descend the hill, through the heather and past the streams, and as my heart returns to normal after the grouse episodes, I spot a large object moving in the sky. It is often said that if you see a big raptor overhead and you can't work out whether it is an eagle or a buzzard, it'll be the latter.

But there is no mistaking this bird. It is huge and soars on winds that are being whipped up by the ridge to the west. It's spotted something, I think, but seems hesitant to do anything about it. After a few minutes it turns towards the wind, without as much as beating or flapping a wing, and disappears over the crest of the hill. No one who has seen a golden eagle can ever forget it. More than any other living thing in Scotland, the red deer and wildcat included, it is the embodiment of the wild. In many churches, lecterns are fashioned from wood in the shape of an eagle. From ancient times it was believed that the bird could stare straight at the sun. Christians, with their apparent ability to gaze fearlessly at the revelation of the divine word, adopted the eagle as a metaphor for their faith. Roman legions used them on their standards. Eagles have long been admired by us.

I reach the bottom of the slope and pick up the path, at last. After several hours of traversing uneven ground, it is good to once again be on a semblance of something that is kinder to the limbs. The wind is now at my back as well. Only a few miles left to get back to the car. I walk for a few minutes. Out of the corner of my eye, off to the side of the path and semi-buried in deep peat, the bleached remains of ancient pine tree stumps and roots lie. Some may be thousands of years old. With their white root systems spreading out in all directions they look like massive octopus carcasses that have been petrified by staring at some Greek monster or other. Back in a time when this land was warmer and less wet, the trees thrived here. But Scots pines dislike waterlogged ground, so eventually these ones succumbed to a changing climate and no shoots could germinate from their spread-out cones. Once the trees died and had fallen, the peat overtook the stumps and buried them. Only now, as we live through a comparatively drier spell, the peat diminishes, and the stumps re-emerge.

I look up, high on the side of the hill to my east. The descendants of these bleached skeletons stand on better drained ground. They sprout seemingly from between large boulders. The terrain around them is rocky, enough to deter the sheep and deer from feeding on them. Small pines grow at their parents' feet. A good sign. On the second-from-last tree, jutting out at an angle, a huge pile of twigs and sticks is visible through the binoculars. It can only be an eagle's nest. An eyrie. Nothing else could be so large. The tree must be over 200 years old. Its thick trunk has withstood scores of years of weather but shows no sign of being defeated. Perhaps the eagle's nest is ancient as well? Some nests can be used for generations by the bird.

The canopy of green contrasts with the sun-bathed pink granite of late afternoon, and as I round a corner I hear the whoosh of the river as it tumbles off a small ledge. A dipper, keen to exploit the turbulent, fast-flowing water, sits on a rock by the riverbank, bobbing up and down. It springs from its perch and disappears into the noisy cauldron of water. After emerging a few seconds later, I marvel at how the water drops off its down as though it were coated in wax. The white bib it wears stands out against the dark water. Luckily, I see two red squirrels corkscrewing around the trunk of a nearby large pine tree that sits close to the river's edge. They seem to be playing. Another treat on a trip that has already provided plenty.

As I cover the final few hundred yards on brittle gravel, the sun has just set behind the large hill to the west. The taller of the hills to my east are still bathed in its glow, but in the river valley it is already getting darker. At this time of the year the temperature drops quickly once the sun has departed. I can see my breath by now and a frost is sure to form. The gravelly path crunches hard as I arrive at the car. Then, silence. A few birds chirp somewhere close. The forest in the distance catches the last of the light, its canopy emerald-coloured by the horizontal rays of the sun. The snow I had been to many hours ago is now a memory. So much of the day was spent walking to and from it. The snow is the reason for the trip, but it would be so much less of a day without the memorable, if brief, sideshows that go on all around.

* * *

The devotion

Although this book is almost exclusively about snow and the hills where it reposes, there is more to seeking it out than just the act itself. In going to these shrines of winter, walkers spend an inordinate amount of time in a semi-wilderness environment that they would not normally visit. That being so, they are disproportionately exposed to – and influenced by – the unusual and thought-provoking things they see, hear and feel around them. For many this may be a profound experience; for others not. Lots of folk are content just to walk to the top of a hill and back down again. A day's exercise such as this is a fine thing on its own and need be examined no further. There are some of us, however, who struggle to leave the memory of it behind so easily.

Hills and mountains speak to people like us in a way that few other things are capable of. They invoke a primeval memory within. In bygone days Palaeolithic and Neolithic man *generally* avoided these lofty places. Little in the way of food could be gathered there, and the weather was often far less reliable than at ground level. Therefore, mountains were, in the main, to be admired for their beauty but, for their unpredictability, kept at arm's length. This mindset remains with us today, I submit, through a process of atavism. A good deal of the public avoids going to the high places, in part because of an association with danger. (It may be postulated that humans' innate fear of being in forests at night is similarly atavistic. It is no coincidence that writers and filmmakers through the ages have used wolves in forests as a plot device to invoke in the reader or viewer a primordial fear.) It cannot, surely, be any kind of accident that some of the language certain people use to describe climbing hills or mountains is so adversarial in its tone. Peaks are not climbed but *conquered*, as if they were foes to be triumphed over. Hills are *bagged* as though shot for sport, like grouse in August. Walkers talk of a *plan of attack* when contemplating the best route up the hill. Part of the reward for many walkers is the overcoming of this physical and psychological barrier. Reaching the top of a high hill can become addictive – an obsession, almost.

As much as some walkers seek to *conquer* a hill, I believe that in some measure it is a conquering or overcoming of *themselves* that is the foundation for their desire. But a hill can never be *conquered* by a walker. For a while he or she may stand on its summit and gaze out to the view that will have changed little since the last Ice Age, but the moment lasts only fleetingly. When that walker goes home and to bed, as they will do that night and every night for the rest of their lives, the hill will endure, even after the memory of the *conqueror* is long forgotten. The notion that one can *conquer* or *bag* a hill is an anthropocentric view, short on humility.

For some others, the act of going on to the hill is akin to devotion. These people are, perhaps, less concerned by the act of ticking hills off a list. They are content just to climb a hill – any hill – for no other reason than its own sake.

Church attendance figures in the UK may have declined dramatically in the last fifty years, but our desire for some form of spiritual fulfilment has not diminished at all. Mankind, uniquely, has evolved with a divine curiosity. Since forever, all over the world, *Homo sapiens* has looked around the land it occupies to imbibe meaning from it. But not just the land. Gods of the sky that represent the sun, moon and stars were common in many countries for thousands of years. Tree deities, water nymphs, mountain gods: these historical attempts to understand and give context to that which we are surrounded by were invented by our ancestors to be suitably reverential deities. Moreover, our devotional pull was so strong that we built places where it could be better practised. Whether Neolithic man's erection of stone circles, Ancient Egypt's construction of the pyramids, or the creation of the great Gothic cathedrals of medieval France, our yearning for answers gave rise to us assembling not only a set of belief systems, but huge physical structures to validate and propagate them. Do those without religious beliefs worship the mountains and hills in the same way? Are the people who hitherto would have gone to church now adopting a different form of service when they walk up a hill as opposed to kneeling in a pew?

Humanists and others who evangelise for the removal of religious studies in schools imagine that children freed from its shackles will grow into adulthood without passing on any inherited spiritual beliefs to their own offspring and, so, eventually, it will be weeded from our society entirely, possibly within a generation or two. From this religious extinction, they suppose, a new, fact-based humanity will emerge, with science at its heart and reason as its God. I believe this to be fundamentally faulty thinking. What they are displaying is faith, not scientific reality. A moment's reflection on the history and nature of our species shows we are incapable of relinquishing the need for a sense of mystery or spiritual fulfilment.

Even if science does work out the genesis of the tiny particles that life or the stars are constructed from, people will continue to look for answers and a sense of the divine. Not necessarily in organised religion – which is just a formalised version of devotion – but also in the tangible, everyday terrain that surrounds them. No scientific theory or proof can eradicate this.

Few if any of us now believe in Olympian gods atop mountains, or water nymphs that leap up and down streams. Plenty of folk, however, still look to the outdoors to quench their spiritual thirst. Though many of these people may not necessarily be religious, they are simply acting on a profoundly devotional impulse that is as old and strong as any of the major religious tenets that steer others in the direction of the mosque, church or synagogue. There can be few human needs whose roots run so deep. They will be part of our DNA for a long while yet.

* * *

The island

If one looks at a topographical map of Britain, several features are obvious. The clearest of these is that the landmass itself is homogenous. From Dartmoor to Torridon there is an irrefutable interconnectedness. Several other features are distinct too. The south and east parts of England are largely flat, having only minor hill ranges to ruffle the otherwise smooth and contourless landscape. Wales and Scotland are the opposite, being predominantly mountainous but pocketed with fertile land along river courses and in the lee of high hills. (The Wye Valley in Wales and the Findhorn in Scotland spring to mind.) The hills of northern England rise near Manchester and extend in an almost unbroken chain, via the Pennines and Southern Uplands, to within striking distance of Edinburgh.

One of the consequences of my long love affair with maps was that, from a very early age, I didn't really understand the political concept of borders. I could tell easily enough where county and country borders lay, but they didn't really mean a great deal to me. The national borders on an Ordnance Survey map were so small and insignificant they barely registered. I was interested in natural features: where rivers originated, cliffs, standing stones, chambered cairns and old places with Gothic writing. Arbitrary lines on a map that were man-made and, in any case, modern, didn't interest me. Streams and rivers crossed borders uninhibited, such as one of northern England's rivers, the Tyne, whose ultimate root lies in Scotland.[2] Hills lay in two countries at once. Place names reflected groups of people, not which country or county they resided in.

Not that I had the necessary critical faculties to understand any of the significance of these observations at the time, of course. Only through subsequent reflection did I come to understand that part of my philosophical and political outlook has been shaped by an interest and love of physical geography.

The very northernmost hills in England are indistinguishable from those in southern Scotland to the extent that few people could tell the difference. The Lammermuirs in south-east Scotland would sit quite happily on the Welsh Marches and no one would think them out of place. The Lake District and the Trossachs are cut from so similar a cloth as to be hard for all but a resident of either to tell them apart.

There is a *Britishness* to all these landscapes that is immediately recognisable to me. I say this without a trace of jingoism or chest-beating patriotism. I believe in neither. My sense of what constitutes 'Britishness' is not rooted in a confected, man-made idea of nationhood. (I recoil at flag-waving in general and don't buy into loyalties being dependent on what side of a line on the map one lives on.) The Britishness I am thinking of is purely a convenient term with which to describe a landscape that is wrapped up in its, amongst other things, geographical, zoological and

---

2   The Kielder Water, which is the main tributary of the River North Tyne, has its origin
     on Hartshorn Pike in Scotland, whereupon it is called the 'Green Needle'.

geological history. (The name Britain, with reference to the island containing Wales, England and Scotland, has been with us in some form since Ancient Greek times.) Britain will seldom be mistaken for anywhere else. There are, certainly, times when it can look more continental (as when the Highlands are blanketed in thick snow, or a Cornish beach in summer looks like a Greek island). In the main, though, our hills lack the height of the great world ranges such as the Alps, Pyrenees, Rockies, Andes, Himalaya or even the Atlas Mountains. Though they may not possess the magnificence of these splendid ranges, they have a character way out of all proportion to their modest height.

Walking in the Lake District about ten years ago, I passed a couple of what looked like experienced hillwalkers, probably in their mid-forties. Being in late May, I was heading for Great End, a hill that falls just short of 3,000 feet, but which carries long-lasting snow in most years. The couple were from Bolton. Neither they nor I seemed to be in a particular hurry to go anywhere, as the day was hot and time was no barrier. As we talked, as hillwalkers who are not in a rush are apt to do, I could tell they were seasoned outdoors folk. We discussed some of what we considered to be the finest places in Britain. They spoke fondly of Torridon in Scotland, as well as Dartmoor in the South West of England. We spoke for about ten to fifteen minutes before going on our respective ways. As they hopped down the flagstone path back towards their car, something very odd happened. Just for a minute, I literally had to do a double take as to where in the country I stood and where I was going to. In the short space of time I had been speaking with the couple I had lost the sense of my location and which hills I stood among. This moment passed within a few seconds, of course, but it was a genuinely peculiar feeling. That night, as I sipped on a drink and pondered the day's success (I found some snow on Great End as hoped), my thoughts looped back to the instant after I chatted with the couple from Bolton.

In that moment, standing there in the pleasant afternoon warmth, it really didn't matter *where* I was. In that few-second gap I could have been anywhere in upland Britain. There was something remarkably familiar about it all. So accustomed was I to the different smells, vegetation, rocks, humidity, temperature, streams, trees, walls, noises and wildlife of so much of upland Britain that it all became interchangeable. The uplands of Britain are connected by being on the same land mass, but there is more to it than that. Something intangible, perhaps, but no less real to me for that.

My feelings about this have, over time, been reinforced because of where I have lived. For a number of years, I dwelt on the small Scottish island of Iona, just off the west coast of Mull – itself an island near Oban. What became patently obvious to me whilst living there was the sense that the islands of the Inner and Outer Hebrides were their own country. Whenever I left Iona to visit family or friends on the mainland, it felt like I was travelling to a different nation; the same being true when coming home. More than the journey itself, there was – and is – something *different* about the islands. Something un-Scottish, un-British. Just *island-ish*. I have an enduring and deep love for Iona and Mull, and might still be living there yet if circumstances had been different, but they and other islands are, to me, apart. Maybe it is to do with the act of crossing water acting as a gateway to somewhere else. An invisible threshold, maybe. I'm sure that is part of it, but by no means all. Some Orkney and Shetland folks say much the same thing.

Britain, the island, has a range of upland areas through which a thread runs. This thread is made up of and spun from the strands each of them have in common, such as the flora and fauna. As far as I am concerned there is no difference between Exmoor and the Pentlands. They are all equal parts in the wonderfully and colourfully varied island we live on.

* * *

The realisation

Most of the walks I undertake, either as part of my research or in general, are done alone. There are two very good reasons for this. The first reason is that often the places I go to for snow research are not easy to reach, as will be amply clear by now. I generally do not wish to inflict the experience on people who are unused to difficult conditions. Even sea-soned hillwalkers and mountaineers that I've been with voice surprise at just how tricky the terrain can be. I confess that, within reason, I enjoy walking on such ground. I don't know if this is partly masochistic or what. All I know is that the more challenging the terrain, the more fulfilling I find it. That's part of it. The second reason I tend to walk alone is, on reflection, less complex.

As a child, maps fascinated me. I couldn't get enough of them. Whether it was old AA road atlases or second-hand Ordnance Survey maps it really didn't matter. I pored over every one of them, intrigued at place names, rivers, hills, ancient monuments and everything else in between. It mystified me why no one I knew really shared this passion. Maps were my friends and therefore precious to me. In geography classes I excelled at anything to do with them. One of my several mediocre teachers over the years used to eye me nervously if anyone in class asked a question that had any cartographical connotations, knowing that I'd pounce on any form of slip-up.

One of the consequences of this passion for maps was a realisation that I was *alone* in it. I understood quite quickly that it was something else I seemed to like that none of my peers did. It's not as if I was a loner, either. I was a socially gregarious boy when young. I enjoyed football and other sports. I ran fast and enjoyed being boisterous. But I soon realised that the things I found truly meaningful and which had any form of inner worth were to be found by doing them alone. Nobody else *got it* the way I did, as it seemed to me. As a teenager I used to lie in a dark room listening over and over to the same piece of music for hours, not wanting to share it because I felt embarrassed and I thought nobody else would understand. I would read my atlases and maps alone and not tell anyone, for the same reason.

When I got older and started venturing out into the hills, or just walking some place in general, I found that I enjoyed it more when I was alone. I didn't have any distractions, or people to hold me up. I could come and go where I pleased, whenever I liked. The story remains pretty much the same right up until today. But that is not to say that great trips are done exclusively when one is alone. Some of the very best and most memorable days have been with others. Good companionship on the hill is a precious thing, especially sharing a special moment with someone who understands what is unfolding around them.

I suppose, then, that when I walk to a fast-disappearing patch of snow by myself part of it is me reminding myself of being young. The boy who found interest in the things that other people didn't; the nine-year-old who looked across to Ben Lomond that day in May 1983 and saw a small drift sitting alone. Maybe, even, by seeking out the last drift of solitary snow, what I am really trying to do is revisit a part of my childhood that I've never really been able to get over or come to terms with.

But, equally, it might just be that I'm a hopeless geek.

epilogue

The A9 and the canine

The first fall of snow is not only an event, it is a magical event. You go to bed in one kind of a world and wake up in another quite different, and if this is not enchantment then where is it to be found?

– J.B. Priestley[1]

Quite a few years back, when I had posted something or other on Twitter about the depth of snow on a Highlands hill, I started to correspond with another Twitter user. It transpired that, like me, he was an avid hillwalker and very keen specifically on winter treks. His desire to get on to the hill during that season was, to use his own subsequent words, 'Because snow makes the world feel very different. It makes everything wilder, far more dramatic, perhaps even with an edge of danger and risk. It covers up all man-made things. It makes a noisy world quieter. You feel totally lost in nature when out in the snow.'

After some to-ing and fro-ing with tentative suggestions to meet up, we aligned our calendars and arranged to go for a snowy, midweek sprauchle on Ben Lawers, a huge hill reaching almost to 4,000 feet, right in the heart of the Southern Highlands. The weather on the day we met

---

1   Priestley, J.B. (1928), *Apes and Angels: A book of essays*, London: Methuen & Co Ltd.

in the car park was, at best, patchy. Low cloud swirled all around us, with mist threatening to reduce the walk to a semi-blind stumble. But, still, there was plenty of snow to go at and the forecast suggested that an inversion was possible.[2] For their beauty and transformative effects on the landscape, inversions are highly prized by walkers in winter and spring. Emerging from a cold 'pea-souper' of a day into brilliant, relatively warm sunshine and clear blue skies is one of the great joys of hillwalking. It lifts the spirits and provides views that can, literally, take the breath away. Anyway, when we left the car park that morning there were no guarantees on the weather, only the promise of snow. That was enough.

On the walk that day with us was a dog: a chocolate Labrador, called Olive. With her unquenchable enthusiasm and seemingly boundless energy, Olive made a memorable companion. I could have had no idea at the time that this unassuming, placid and hungry hound would go on to become – it is no exaggeration to say – internationally famous as one half of the Olive and Mabel Labrador pairing that quite justifiably deserves the handle of 'internet sensation'. Olive's (and Mabel's) owner, and my walking buddy that day, was, of course, BBC commentator Andrew Cotter. During the 2020 lockdown, Andrew put his energies and, frankly, abundant Covid-related spare time into making a series of videos that went viral, and which have since been viewed literally tens of millions of times. But on that quiet, midweek April day at Ben Lawers, it was just Olive, Andrew and me. Mabel was only the proverbial glint in the eye of her doggy parents. But what a day we had. Ben Lawers, resplendent in miles of unbroken spring snow, was an absolute treat. The inversion that we hoped for duly arrived halfway up the hill, exposing only the top 1,000 feet of the surrounding hills in their naked glory. As we climbed higher still, we saw the thick cloud hide the cols that conjoined the neighbouring hills, giving them the appearance of lumps of ancient rock emerging from a grey sea. Nothing could have been finer.

---

2   An inversion is when low cloud sits in the valley below, due to the cold air sinking to the bottom of it. But as height is gained, the temperature rises slightly, so that one bursts through the cloud and looks down on top of it, not unlike when an aeroplane emerges from the gloom of a wet and miserable day into clear, unbroken blue skies.

Watching the interaction between Olive and Andrew that day was fascinating. As someone who didn't grow up in a dog household and had never owned one, I was always puzzled as to why so many people invested so much emotional (not to mention financial) capital in them. To me they were nice playthings, but they seemed to be awfully hard work – as well as being quite a tie. I was never a fan of the *wet dog* smell either. Also, when I was young I saw a couple of dog-attacks on children, so I had a bit of reticence about their behaviour. However, this was the first time I'd ever been in the company of a dog for such a long snow walk. And, I must confess, by the end of it I was smitten. Watching Olive that day, bounding around with absolute, unadulterated joy, was a real tonic. Glissading, or *bum-sliding* to give it its less technical name, down a steep bank of snow was of particular delight to Olive. She didn't seem to get it. 'Human Is Moving Without Moving His Legs' seemed to flash through her head – so far as anything can flash through a Labrador's head. This activity elicited a lot of barking and jumping up and down. All of this, of course, was met with sheer adoration in Andrew's expressions and actions. Many chicken-y treats were dispensed that day, each of them eaten more quickly than I thought physically possible. For a first walk, it was definitely memorable.

It can be a risk meeting someone in the flesh on the strength of their social media profile and limited interactions you might have had online. Fortunately, though, Andrew and I seemed to have a good deal of out-doors interests in common, and the day passed all too quickly amid stun-ning scenery, steep banks of pristine spring snow, and dog-related fun.

Over the next couple of years, we met up regularly in winter and would go for midweek snow walks in out-of-the-way places, with Andrew always keen to avoid the honey-pot spots that tended to draw the crowds. In terms of fitness and passion for the hills we made good companions. There was, and remains, a tacit, unspoken respect for each other's abilities and capacity to get up at 4.30 a.m. to drive 200 miles just to grab the opportunity of a good day out. The latter is absolutely essential if one is to seize the very short weather window opportunities that Scotland offers in winter.

* * *

I think I can remember most of the walks that I have done over the years. Of course, I don't recall all the details of everything seen or done on these days out, but there is usually something that happens which leaves an imprint on the memory. An eagle sighting, maybe, or the remains of an avalanche. It might only be something fleeting, but that can be enough. One of the very best recent trips was in January 2018, and it was the first time I had the pleasure of meeting Andrew's new Labrador, Mabel. On this particular day, however, the amount of memories experienced means that this is likely to last longer than, and be remembered just as fondly as, any walk I've done for quite a long while.

We had arranged to meet up early in the morning, as usual. The weather promised to be as good as you could ever hope for during a Scottish winter. As I drove north that morning the pitch-black sky was pierced by the pinlight of a million stars. The thermometer on the car showed sub-zero temperatures, and the snow that lay on the ground was covered in a hoary, deep frost. As I crossed the northern Perthshire boundary into Highland district the sun started to rise, turning the inky-black sky slowly to the colour it would remain all day: a glorious, deep metallic blue. The sun's low winter trajectory meant that it took a while for it to climb above the low hills to the east. But rise it did, just as I pulled into the car park to meet Andrew, who had arrived not more than a few minutes before. As I stopped beside his car, I noticed the silhouette of two dogs in the boot, both with tails wagging so hard that they might shatter the windows. As he popped the tailgate, the familiar figure of Olive jumped out first, closely followed by a blonde Labrador. This was Mabel. Perhaps worried that one of them might be hit by a lorry on this busy Highland road, Andrew wanted to get going as quickly as possible. Apart from anything, days were short at this time of the year, and we had a big, ten-mile round ahead of us in deep snow.

The four hills ahead of us were coated in the most beautiful white powdery snow. Thankfully, though, a strong crust formed on top, meaning that, once a bit of height was gained, it took our weight, thereby avoiding the need to do the incredibly tiresome and tiring job of breaking trail. The higher we got, the less vegetation we saw. Even the long grass and tall heather were invisible beneath this huge volume of snow. All the while Olive and Mabel were going absolutely bananas playing in the snow. If Andrew or I stopped for a breather or to take a picture, creating a small gap between us, the dogs would expend untold energy darting back and forth almost as if to make sure we knew that they were looking out for us.

I recall the futility of Mabel trying to chase a mountain hare up the side of a hill. This latter animal, supremely adapted over thousands of years to evade the stealthiest and quickest pursuers, made Mabel look like a malfunctioning robot. Never have I seen such a one-sided contest. By the time Mabel had reached the bottom of the slope where the hare had been, it had long disappeared over the top and was probably half a mile distant. The older, and infinitely wiser, Olive watched on impassively and incredulously.

Our walk in, along the Perthshire–Highland boundary, was done in the shadow of the magnificent hill called the Boar of Badenoch. It gazed lustily across the floor of the glen to the Sow of Atholl. These two fine hills of old marked, as their names suggest, the outer edge of Atholl and the start of Badenoch: two ancient regions of Highland Scotland.

In the end, Mabel's pursuit of hare, grouse, ptarmigan, and anything else that moved that day, meant that by the last summit she was whimpering and tired. At one point she tried to fashion herself a little nest out of the snow. Once again, Olive, who had seen it all before, stood stoic and impassive. As with Ben Lawers, I could really see how the presence of dogs could add to the experience of being on snow-capped peaks. The sheer unbridled joy and pleasure of the dogs in this environment was heart-warming. I was beginning to come round to the idea that perhaps I had, hitherto, been too hasty when dismissing them as more hassle than they were worth, irrespective of their loveliness. Maybe the time is nearing when I get one of my own.

Apart from the dogs, what made this trip so memorable was a combination of weather, hills, and the deep snow. Not once in the entirety of the walk did the sun threaten to disappear behind a cloud.

The deep blue winter sky could not have contrasted more starkly with the gleaming white of the snow. The views from the summits, across much of the Central Highlands, were unsurpassable. As pretty as these hills are in summer, they cannot compete when covered with snow. As I say to most people when out walking, snow makes everything better.

General terminology and Gaelic pronunciation

Place names in the Highlands of Scotland are overwhelmingly of Gaelic origin. This is *Scottish Gaelic* as opposed to *Irish Gaelic*. The former is pronounced as 'GAAH-lick', the latter 'GAY-lick'. It's an important distinction that native speakers can become irked at.

It is a long-held and generally established fact that Scottish Gaelic was brought to these shores via Irish settlers around AD 500.[1] The language spread eastward for the next few hundred years until it eventually became the spoken tongue for much of the country. In the twelfth century, virtually the entire population spoke it, save for parts of Lowland Scotland (including Lothian and Caithness), who spoke variants of Scots and Norse.

As is normally the case throughout history with languages in a time of social and political instability, Gaelic's dominance began to slip over the centuries. Lowland Scots, now spoken by the landed class and increasingly at the royal court, started to penetrate westward through the glens and river valleys. The drop in numbers of people speaking Gaelic accelerated after the 1745 Jacobite uprising, when it was proscribed by the British government. Children were censured in schools from

---

1   Jones, Charles (1997), *The Edinburgh History of the Scots Language*, Edinburgh University Press.

speaking it up until relatively recent times, which accounted for further significant falls in numbers. (It was seen as a backward language by some.)

Records from the last 250 years paint a sobering picture. In 1755, twenty-three per cent of the Scottish population spoke Gaelic (290,000 speakers). In 2011, it was just 57,602 (1.1 per cent of the total population).[2] Though substantial efforts are being made to revive the language, both in cities and in the country, as a spoken, community language it is to a very significant extent restricted to the Outer Hebrides. Some parts of northern Skye and other Inner Hebrides have small, isolated communities, but with over half the native speakers living in Glasgow, the future for the language as an everyday one is very unclear.

It is true that some of the very earliest strata of names, such as those for rivers like the Dee, Clyde or the Tay, pre-date even Gaelic and are probably thousands of years old. Most names though, especially hills, are Gaelic in origin.

Pronunciation

Like the dialects of England (Brummie, Scouse, Geordie, Yorkshire, East Anglia, etc.), Gaelic is not a homogeneous language. Though words may be written the same for places in Mull as they are for Perthshire, they often have different pronunciations. Trying to negotiate this is a minefield, and to their credit some websites make attempts.[3] However, often fine institutions and well-meaning websites do make mistakes. This is no truer than in the Cairngorms, where local pronunciation is quite different to other parts of the country.

Throughout the book I have attempted to give rough translations and meanings wherever possible. The list below is more authoritative, giving the names of the main hills contained therein. The name of the hill comes first, in bold, followed by an approximate sound in Scots/English, indicated by italics. Emphasis of syllables is given in upper case. Then follows the meaning, with an explanatory note if necessary.

---

2   MacAulay, Donald (1992), *The Celtic Languages*, Cambridge University Press.
3   https://getoutside.ordnancesurvey.co.uk/guides/the-gaelic-origins-of-place-names-in-britain/ [accessed 13 May 2021].

Aonach Beag *eunuch BAKE* – 'little ridged hill'. The *ao* sound in Gaelic has no English equivalent. The best way of mimicking it is to try and say *oo* without rounding your lips.

Aonach Mòr *eunuch MORE* – 'big ridged hill'.

Beinn a' Bhuird *been-a-board* – 'table hill'. The 'a' is very short, and hardly pronounced. It indicates a euphonic (for ease of pronunciation), not the article. The designation 'Bhuird' is incorrect. Older folk pronounce the hill, approximately, as *board*. Recently *voord* or even *voorsht* have become common.

Beinn Bhrotain *been VROHtan* – 'hill of the mastiff'.

Ben Avon *ben AAHn* – meaning unsure. Possibly from *athfhinn* ('very bright one'), or from *abhainn* ('river').

Ben Macdui *ben macDOOi* – 'hill of Macduff'. The highest hill in the Cairngorms and the second highest in the UK.

Ben Nevis *ben NEVis* – meaning unknown. An ancient name, possibly pre-Gaelic. One interpretation is 'hill with its head in the clouds', though this is disputed.

Braeriach *bray REEach* – 'the brindled upland'. The first syllable is generally pronounced as 'bray', but – properly – is 'bry'.

Cairn Gorm *karn GORom* – from *An Carn Gorm*, meaning 'blue hill'. This is the hill that gives its name to the whole range.

Cairn Toul *karn DOWel* – 'hill of the barn' (from *Carn an t-Sabhail*).

Geal-Chàrn *g-YAHL charn* – 'white hill'.

Liathach *LEE-a-huch* – 'the grey one'.

Sgòr an Lochain Uaine *scorn lochan OOan* – 'peak of the green tarn' (small loch). Also known as the Angel's Peak, a name given by Alexander Copland as a counterbalance to the nearby 'Devil's Point'.

Sgòr Gaoith *scor GOOee* – 'peak of wind'.

Sgùrr na Lapaich *scoor na LAHpeech* – 'hill of the bog'.

Get involved

Though there are now many people who are actively involved in the study of snow patches and snow in general, there is always room for more. The uplands of the UK are huge, with many nooks and crannies where relics of winter can cling on unseen, just waiting to be discovered by a walker, climber, cyclist or caver. To that end, I encourage people who stumble across these old patches to get involved in their reporting. The best way to do this is via two specific pages for Scotland, England and Wales on Facebook:

www.facebook.com/groups/snowpatchesscotland
www.facebook.com/groups/snowpatchesengland
(for England and Wales)

There are now several thousand people on these pages, many of whom are very knowledgeable and keen to engage with new members.

The other medium is Twitter, specifically via my own page. I can be found at @theiaincameron all year posting updates and interesting snow-related pictures. Please feel free to say hello on there if you've enjoyed this book.

Acknowledgements

Not in any particular order, but I would like to thank the following people for helping to bring this book to print:

Laura – for her unceasing and helpful advice on what worked and what didn't when I was trying to knit the story together.

Jenny Brown – my agent (www.jennybrownassociates.com) for her excellent and constructive suggestions and having faith in what I was trying to say.

Vertebrate Publishing – for the belief they showed in taking on something that was outside of their natural comfort zone. To Commissioning Editor Kirsty in particular.

Murdo MacLeod – the master photographer. His photos from various trips he and I have completed in the last few years have helped to bring this subject to a wider audience.

Andrew Cotter – for his permission to allow me a shameless addition of a doggy epilogue.

Finally, to the late Adam Watson. Despite it being over two years since his passing, I cannot help but say to myself, 'I must send that photo to Adam,' when I see one of the Cairngorms covered in snow. His knowledge and passion remain a daily inspiration.